Sustainable Development Goals Series

The **Sustainable Development Goals Series** is Springer Nature's inaugural cross-imprint book series that addresses and supports the United Nations' seventeen Sustainable Development Goals. The series fosters comprehensive research focused on these global targets and endeavours to address some of society's greatest grand challenges. The SDGs are inherently multidisciplinary, and they bring people working across different fields together and working towards a common goal. In this spirit, the Sustainable Development Goals series is the first at Springer Nature to publish books under both the Springer and Palgrave Macmillan imprints, bringing the strengths of our imprints together.

The Sustainable Development Goals Series is organized into eighteen subseries: one subseries based around each of the seventeen respective Sustainable Development Goals, and an eighteenth subseries, "Connecting the Goals," which serves as a home for volumes addressing multiple goals or studying the SDGs as a whole. Each subseries is guided by an expert Subseries Advisor with years or decades of experience studying and addressing core components of their respective Goal.

The SDG Series has a remit as broad as the SDGs themselves, and contributions are welcome from scientists, academics, policymakers, and researchers working in fields related to any of the seventeen goals. If you are interested in contributing a monograph or curated volume to the series, please contact the Publishers: Zachary Romano [Springer; zachary.romano@springer.com] and Rachael Ballard [Palgrave Macmillan; rachael.ballard@palgrave.com].

Eileen Pittaway · Linda Albina Bartolomei

Only Rape! Human Rights and Gender Equality for Refugee Women

From Refugee Camps to the United Nations

palgrave
macmillan

Eileen Pittaway
Forced Migration Research Network
University of New South Wales
Sydney, NSW, Australia

Linda Albina Bartolomei
Forced Migration Research Network
University of New South Wales
Sydney, NSW, Australia

ISSN 2523-3084 ISSN 2523-3092 (electronic)
Sustainable Development Goals Series
ISBN 978-981-19-0915-3 ISBN 978-981-19-0916-0 (eBook)
https://doi.org/10.1007/978-981-19-0916-0

This Palgrave Macmillan imprint is published by the registered company Springer Nature Singapore Pte Ltd.
The registered company address is: 152 Beach Road, #21-01/04 Gateway East, Singapore 189721, Singapore

This book is dedicated to Akhu, a Chin refugee woman who lived in New Delhi, a long-standing colleague and friend, with whom we had the privilege of working for eleven years. Akhu was a refugee and women's rights activist, a key member of Women for Justice which she represented on the board of the Women's League of Burma. She helped thousands of women through her advocacy and training and the safe house she managed. Refusing the opportunity to resettle to the USA, she chose to return to Burma/Myanmar in 2016 and work with her people to rebuild their lives. We had planned to collaborate with her in 2018/2019, but we could not travel to Burma because of the renewed conflict. She was shot and killed by the Burmese military, on the 28 March 2021, while peacefully protesting against the erosion of democracy in the country. Vale Akhu.

Acknowledgements

We acknowledge and pay our deepest respects to the Gundungurra and Darug peoples on whose sovereign and unceded lands we are privileged to live and on which this book was written.

From Linda

To my mother Colleen Bartolomei, ever a tireless advocate for the rights of refugees, who joined us for part of this journey, as a volunteer teacher in the camps in Thailand and in New Delhi, and for all her editorial support over many years. To my father Sergio, who had he still been alive, I would be thanking for his moral support and the whiskies we had shared along the way! Thank you to my many wonderful friends, family and colleagues in Australia, England, Israel and Italy—for meals shared, mountains climbed, rivers and gardens explored, and aperitifs drunk as the sun set over ancient piazzas. Thank you all for always being there to assist, both morally and financially, in response to urgent calls for support from refugee women and their families. Special thanks to Teri and the team at Yeah Hair Baby Marrickville for your sustained financial support over many years for small projects initiated by the refugee women and to meet emergency needs. To my nieces Magenta and Siena, my god-daughter Angelina and to Cliona—thank you for the fun and joy you have added to my life through your many visits, since you were little girls. To Eileen Pittaway, friend, collaborator and inspiring mentor. It has been the greatest of privileges to have joined you on the journey you had begun. I am so very proud of what we have been able to achieve together.

From Eileen

To my spouse Judi Tucker who has put up with me every step of the way, wiping my tears, celebrating successes, editing reports, preparing training materials, making our films, keeping our wonderful home while I have run around the world and come back to be tied to the computer writing reports and papers. I can never thank you enough. To my wonderful daughter Emma, this story is the backdrop to much of her life. She has edited, researched, written and delivered training materials, worked overseas with us as a volunteer and been a fierce critic of my work. I am a lucky mother, Thank you. And my fantastic son-in-law Tashi and my gorgeous grandsons Ashley and Jasper, for being there for me and making life so happy. Over the years so many friends have been there to support me whenever needed. Special thanks to Barbara Blair who in 1990 voluntarily transcribed the over 300 interviews of rape and torture which began this journey. In this last phase of my life, writing this book, special thanks to Jennie Mosely and Patricia Woodley who have supported Judi and I by caring for our dogs whenever needed (which was often) and always making sure we were OK. And our broader group of friends, thank you for the food, the drinks, the laughter and friendship on which we can always depend. Finally, to Linda Bartolomei, co-author, conspirator and wonderful friend. It has been a privilege to work with you. Without you I would have given up years ago.

From Us Both

Special thanks to our readers, critics and editors Colleen Bartolomei, Jennie Mosely, Emma Pittaway, Judi Tucker and Rameen Zahra, who have done an amazing job in guiding us to tighten and improve the text, not be repetitive and to remove surplus commas, and, Capital Letters. All mistakes can be credited to them!

To Velvy Holden, founder of ANCCORW, whose dream this was and Juliana Nkrumah from Ghana who was a prime mover in establishing the NGO. To the staff of the UNSW Centre for Refugee Research, Effie Mitchell, Geraldine Doney and Rebecca Eckert, who were with us for most of the journey, and to Carole Shaw, Kristy Ward, Marcela

Garett, Rochelle Baughan, Olivia Wellesley-Cole, Georgia-Rae Cobon and Deborah Raphael, academic colleagues Eileen Baldry, Kerrie James and Richard Hugman, and the many student interns who joined us on parts of the journey. All worked day and night in their dedication to the refugee women with whom we collaborated. And we do not forget our academic colleagues at UNSW, for their contributions to the work of the Centre for Refugee Research, their willingness to volunteer, self-fund to travel with us and to cover for us so that we could attend meetings overseas. To the many colleagues in UNHCR Geneva and around the world who receive a lot of criticism, but who never get enough recognition for the amazing and difficult work they do. To the many wonderfully supportive staff across various departments in the Australian Government, who have almost continually and generously funded this work for 33 years. To Rosie Scott, who generously provided the funding for the post production of our film, 'Only Rape'. And to our fantastic reviewer, Dr. Lisa Hartley, thank you for the detailed and thoughtful response to the book. We hope that we have done justice to your excellent suggestions for change. And most importantly to the many, many incredible refugee women and refugee men with whom we have had the privilege of collaborating for so many years.

To All the People We Have Named and Not Named

This is an epic story with an amazing cast of characters. We want to name each and every one of you, because without so many amazing people, there would be no story. We would be quietly teaching our courses at university. At first, we started to acknowledge everyone by name, but in places the pages looked more like a list than a story. Our faithful first readers told us to stop, to find some other way to recognise your contributions. We came up with the idea, of naming but a very few in the text, and including others in an endnote to each chapter. We are sure we have missed many of you and we can only apologise in advance. We hope we have not claimed credit for an idea that you had, for something that you did. If so, we are deeply sorry, and please let us know. I can assure you in advance it was not intentional, we have spent a lot of time plumbing the depths of our memories and our notes to ensure that the version of this story we present is as accurate as we can make it. The story has gone on for so long, and at times, we were running so fast our feet barely touched the ground. Thank you for being there with us.

About This Book

This book documents the story of 33 years of research and advocacy for the rights of refugee women and girls and their improved protection from endemic rape, sexual and gender-based violence. It is a journey taken by two academics from the University of New South Wales, Australia, Associate Professor Eileen Pittaway and Dr. Linda Bartolomei. The story is in two stages, the first, chapters one to four, the fight to have rape in situations of conflict, recognised in international law, and what triggered this, is Eileen's story. The second part, chapters five to eleven, is our joint work to have the law implemented. We have developed this book together, a close collaboration, very much in the style in which we have been working together all these years. It is our shared story, and it is written in both of our voices. At the beginning of each chapter, we have indicated who is the narrator and it swings back and forth between the two of us. Each chapter contains a stage in our journey and introduces the theory which we used to analyse, understand and inform the next cycle of research. It is a complex story, and we hope that it is not too confusing. In all the years of research, we have meticulously documented our research findings, consultations with communities, meetings, interviews and our observations. These notes form the basis for much of the story. To avoid bombarding the reader, there are many details and sub-plots left out, and we have referenced relevant reports and papers to guide you if you would like more information.

A Gentle Warning

The book contains accounts of rape, sexual abuse and torture endured by refugee women and girls told in their own voices. They are an integral part of the story we need to tell. A refugee woman who was about to make a presentation at a conference we attended was asked by the facilitator not to give any graphic examples of what had happened to her because it might traumatise some of the audience. She looked her straight in the eye and said 'I am sorry, we survived it, they can hear it. If nobody listens nothing is done'. When we convene events at which refugee women give testimony, we arrange for a counsellor to be present in case anyone becomes very upset.

How We Have Used Boxes in the Text

> In the book, we have put the very confronting stories and content in boxes in *Italics*, so you can skip over them if you are feeling too distressed or uncomfortable. Other boxes, where the text is not in italics, highlight the key points we want to raise in each chapter. If the text in the boxes is a quote, it is referenced as such, otherwise it is our findings and analysis.

What's in a Name? From Refugees to Forcibly Displaced Persons

As the story begins, we refer to refugees as that group of people covered by the mandate of UNHCR, who have fled their homelands because of persecution and are displaced in countries of refuge. However, during the 1990s at the request of the UN Secretary General, UNHCR began to play an increasing role in the protection of some of the growing number of people internally displaced (IDPs) by civil war and disasters (UNHCR 1998). In 2005, there was movement to further formalise a response and the Global Protection Cluster, of which UNHCR was a key member, was established. It was agreed that UNHCR would assume global cluster leadership for protection and co-leadership in the area of camp coordination/management and emergency shelter (UNHCR 2013). At the same time, UNHCR began to take a larger role in relation to the prevention and reduction of statelessness. This increased the number of people who were included in the mandate of UNHCR. Many member states were unhappy with this as the agency was struggling to find the resources to meet the needs of refugees. This group now includes people displaced as a result of climate change and natural disasters. Over the years, those under UNHCR's mandate have been referred to as 'affected populations' and 'people of concern', and more recently, we refer to 'refugees and forcibly displaced people'. It is still a grey area as people granted refugee status fit a very specific legally defined criteria, which is not so clear for other categories of displaced persons. What is important to remember is that they all need and have the right to international protection. Throughout the book, we have endeavoured to use the terminology which was most used at that time in the story.

How Have We Defined and Used the Descriptors *the Global North and the Global South, Developed and Developing Countries*

There are tensions in the use of these terminologies, which have changed in meaning over time and in different contexts. Over the course of this story, the ways in which different parts of the world have been referred to in the context of socio-economic and cultural status has shifted significantly. Each time, the language used is laden with history, and evolving ideological and socio-economic perceptions of countries and regions. When we started this work, the discourse of first world i.e. developed, affluent nations, and third world, i.e. underdeveloped nations, often former colonies of the first world, was still in use. There was a notion of a second world, of quickly developing nations, but this was scarcely used. These terms were value laden, with white supremacy implicit in the discourse. Interestingly, while spell checking this work, my computer suggested that using these terms might imply cultural bias! Over time, these terms were replaced by reference to the global north, which contains over 90% of the worlds manufacturing capacity and earns over 4/5 of the world's income, and the global south, which houses 3/4 of the world population and earns only 1/5 of the world's income. These terms were perceived to be less pejorative. However, the issue is not just a geographic divide, and there are developing countries in the north and developed countries in the south. Instead, the terms global north and global south have come to be used as indicators of socio/economic position in the world of globalisation and development. This terminology and the meaning behind it is highly contested, we could write a book just on this!

In this work we have accepted these terms, and the simplistic notion that part of the function of the global north, as global citizens and as part of the United Nations family, is to provide *Development Aid* to economically disadvantaged countries of the global south. The purpose of development aid is to grow the economies in the global south, both to assist and raise the standard of living in individual countries and as an international good, in particular in terms of the global economy. As countries become more developed, they are referred to as 'countries in transition'.

However -and there always is an 'however', since the end of the second world war, 1947, the world refugee and displaced population has grown, to an estimated 100 million, (UNHCR, May 2022). The majority, but definitely not all of these have been both produced by countries in the global south and been accepted and hosted by other global south nations. Billions of dollars of aid have been spent supporting refugees and the local countries to which they have fled, and this has been labelled *Humanitarian Aid*, the purpose being an immediate response to a humanitarian crisis, a short-term endeavour. What we have seen is that the majority of refugee situations are not short term, many have continued for years, but in many places, the aid response has stayed at the level of a humanitarian response. In the twenty first century there has been a growing realisation that aid to refugees has to be a mix of both. A crisis response in the first instance, followed by a speedy transition to the use of a development aid model wherever possible. This is the way we have used these terms in this book. For more detailed and nuanced analysis of these complex issues we refer you to the plethora of academic articles available online addressing discussion of the Global North and the Global South and the websites of the United Nations Development Program (UNDP) and the Office for the Co-ordination of Humanitarian Affairs (OCHA).

Burma or Myanmar

To clear up a possible misunderstanding before it occurs, when we first worked with refugees from Burma on the Thai Burma Border and in India, they were adamant that they would not call their homeland

Myanmar, as this was a name imposed by the corrupt Government which had forced them to flee. They later accepted the name when the civil war was declared over, and for a brief window of time, it looked as if returning home might be possible. We interchangeably use the name preferred by the refugees with whom we worked, at different times in the story.

Our Place in the Story

The story is our story; it has been our lives for the past three decades. However, it must be recognised that on the long journey, we play a minor role. We are but one cog in an enormous wheel of international advocacy, working alongside many other activists, passionate, committed women and men who have contributed their time, wisdom, knowledge, expertise and dedication to the struggle for human rights for refugees. As white academics, we must take the backseat and acknowledge the courage, determination and skill of the refugee women who have worked so hard and overcome so many challenges to fight for their rights. They are the ones who most deserve recognition and least often get it. Sadly, we fear that this struggle will continue for the next generation of refugees, activists and academics, but the significant gains that have been made at different stages of the journey make it all worthwhile. They prove the power of academic research to contribute to positive change.

Contents

About the Authors

Eileen Pittaway was director of the Centre for Refugee Research, UNSW, and associate professor in the School of Social Sciences and International Studies, coordinating and teaching in the master programs of International Social Development, and Refugees and Forced Migration from 1999 to 2013. She now continues her research activities and involvement with the university in an honorary capacity. The major focus of her work has been the prevention of and response to the rape, sexual abuse and gender-based violence experienced by refugee women, both overseas and following resettlement to Australia. Over the past 24 years, she has worked with refugees, United Nations and humanitarian aid agencies and evaluated projects in 22 different countries. In 2001, Eileen was awarded a Human Rights Medal by the Australian Human Rights and Equal Opportunities Commission for her work with refugee women and children. In 2005, she received a New South Wales Premier's Award for services to refugee education. In 2011, on the centenary of International Women's Day, she was named one of the hundred women who have shaped Australia. In 2012, she was made a member of the Order of Australia for her work with refugees.

Linda Albina Bartolomei is the immediate past director of the University of New South Wales (UNSW) Centre for Refugee Research, a co-convenor of the Forced Migration Research Network (FMRN) and a senior lecturer in social work and international development. Since 2002,

Linda has co-led a series of action research projects exploring the challenges associated with identifying and responding to refugee women and girls at risk in camps and urban settings. This involved research in multiple sites across Africa, Asia and the Middle East, and in Australia. The results had significant impact on research and policy and were instrumental in the introduction of the United Nations High Commissioner for Refugees (UNHCR) Conclusion on the Protection of Women and Girls at Risk in 2006. In 2013, she was awarded the UNSW Community Engagement award for her leadership of an innovative practice-based research project in New Delhi, India. The Refugee Community Development Project (RCDP) was a refugee community-led project implemented in partnership with the Afghan and Somali Refugee communities. Over the last three years, she has worked with UNHCR Geneva and a team of women from refugee backgrounds to advocate for and monitor the inclusion of refugee women and girls' protection needs and rights in the new Global Compact on Refugees (GCR). She is currently co-leading a multi-year project in four countries in the Asia-Pacific to support the implementation and monitoring of the commitments to refugee women and girls in the GCR.

Abbreviations

ACOA	Australian Council for Overseas Aid
AGD	Age, Gender and Diversity
AMES	AMES Australia
ANCCORW	Australian National Consultative Committee on Refugee Women
ANCORW	Australian National Committee on Refugee Women
APWLD	Asia-Pacific Women in Law and Development
APWW	Asia-Pacific Women's Watch on Human Rights
ARC	Australian Research Council
ARRA	Australian Refugee Rights Alliance
ATCR	Annual Tripartite Consultations on Resettlement
AUSTCARE	Australian-based, aid organisation now Action Aid Australia
AWAU	African Women's Advocacy Unit
AWHRC	Asian Women's Human Rights Council
B+10	Beijing Plus Ten
B+5	Beijing Plus Five
BBC	British Broadcasting Corporation
BIMPR	Bureau of Immigration, Multicultural and Population Research
BIR	Bureau of Immigration Research
BPFA	Beijing Platform for Action
CAAIP	Committee to Advise on Australia's Immigration Policies
CAPOW!	The Coalition of Participating Organisations of Women
CATAW	The Coalition for Assisting Tsunami Affected Women
CBO	Community-Based Organisation
CEDAW	The Convention on the Elimination of all Forms of Discrimination against Women
CNN	Cable News Network

CONGO	Committee of International Non-Government Organisations
CRR	Centre for Refugee Research
CRRF	Comprehensive Refugee Response Framework
CSW	Commission for the Status of Women
DAW	Division for the Advancement of Women
DFAT	Department of Foreign Affairs and Trade (Australia)
DIAC	Department of Immigration and Citizenship
DILGEA	Department of Immigration, Local Government and Ethnic Affairs
DIMA	Department of Immigration & Multicultural Affairs (now DIAC)
DRC	Democratic Republic of the Congo
ECOSOC	Economic and Social Council
ESCAP	Economic and Social Council of the Asia-Pacific
FMRN	Forced Migration Research Network
GAD	Gender and Development
GCM	Global Compact on Safe, Orderly and Regular Migration
GCR	Global Compact on Refugees
GIA	Grant in Aid
HCD	High Commissioner's Dialogue (UNHCR)
HRIT	Heightened Risk Identification Tool
HRW	Human Rights Watch
ICC	International Criminal Court
ICMC	International Catholic Migration Commission
ICTR	International Criminal Tribunal for Rwanda
ICTY	International Criminal Tribunal for the former Yugoslavia
ICVA	International Council of Voluntary Agencies
IHRC	Indonesian Human Rights Commission
IRO	International Refugee Organisation
JFFT	Joint Fact-Finding Team
JRS	Jesuit Refugee Service
KnED	Karenni Education Department
KNU	Karen National Union
KNWO	Karenni Women's Organisation
KRC	Karen Refugee Committee
KWO	Karen Women's Organisation
LGBTIQ+	Lesbian, Gay, Bisexual, Transgender, Intersex and Queer
MAP	Migrant Action Program
MRC	Migrant Resource Centre
NGO	Non-Government Organisation
NYD	New York Declaration
ODP	Orderly Departure Program
OSW	Office for the Status of Women
PHAA	Public Health Association of Australia

RCK	Refugee Consortium of Kenya
RCOA	Refugee Council of Australia
RRWG	Refugee Resettlement Working Group
RSD	Refugee Status Determination
SBS	Special Broadcasting Service
SGBV	Sexual and Gender-Based Violence
SPDC	State Peace and Development Council (Burma)
SSI	Settlement Services International
STARTTS	Service for the Treatment and rehabilitations of Torture and Trauma Survivors
SWAN	Shan Women's Action Network
TAFE	Technical and Further Education
TMSG	Townsville Multicultural Support Group
TPV	Temporary Protection Visa
TRuK	Tim Relawan untuk Kemanusiaan (Volunteer Team for Humanity)
UDHR	Universal Declaration on Human Rights
UK	The United Kingdom
UN	United Nations
UNFPA	United Nations Population Fund
UNGA	United Nations General Assembly
UNHCR	United Nations High Commissioner for Refugees
UNHRC	Human Rights Council was Commission
UNSC	United Nations Security Council
UNSW	University of New South Wales
URSP	University Research Support Program
USA	The United States of America
USCRI	United States Committee for Refugees and Immigrants
WaR	Women at Risk
WCAR	World Conference Against Racism
WCRWC	Women's Commission for Refugee Women and Children now WRC
WEL	Women's Electoral Lobby
WFP	World Food Program
WILPF	Women's International League for Peace and Freedom
WLB	Women's League of Burma

List of Figures

CHAPTER 1

Prologue: Why This Story Has to Be Told—A Small Cog in a Large Wheel of International Advocacy

Narrated by Eileen Pittaway

They rape us at borders, they rape us when we go for food, they rape us when we go for water. If we want to leave the camp we have to sleep with the guards; they beat as a warning not to tell. When will it stop? (Congolese refugee woman 2002)

All my sisters, my mother, my friends - all the women have been raped. The military, they rape us. When we try to cross borders they rape us, when we go for water they rape us, when we go for food they rape us, when we go to the bathroom, they rape us. The police, they rape us. Our life is rape. (Rohingya refugee women 2019)

In struggling to find a way to present the many jigsaw puzzle pieces necessary to fully understand this story, we were very pleased to stumble upon this definition of a prologue.

> *A prologue is a piece of writing found at the beginning of a literary work, before the first chapter and separate from the main story. It introduces important information, such as background details or characters that have some*

E. Pittaway and L. A. Bartolomei,
Only Rape! Human Rights and Gender Equality for Refugee Women,
Sustainable Development Goals Series,
https://doi.org/10.1007/978-981-19-0916-0_1

connection to the main story, but whose relevance is not immediately obvious. (Masterclass, 2021)

This prologue is a map to guide you along the way.

An anatomy of rape in conflict – the unspeakable truth!
'Rape of refugee women and girls is so common it has become normalised' (Male psychologist, Malaysia 2019).

Rape occurs at all stages of the refugee journey. It is often part of persecution in homes and villages, as a strategy of conflict, in flight, at borders and in refugee sites. It is perpetrated by military, guards, militia, police, males from host communities and males from refugee communities, sometimes by humanitarian workers. It is an exercise of power and subjugation, not only of women, but also of their families and communities. Women are raped in front of husbands, fathers and children. The impacts are profound. Many abused women bear children of rape. Young girls die because they are too young to bear the children conceived from rape. Some women are so stigmatised that they are banished from their communities. Men are shamed because they cannot protect women and girls, and whole communities suffer collective guilt. It occurs in all aspects of their lives and cross cuts all the areas of the protection they should receive from the international community. Many refugee women are forced to sell sex to feed themselves, their children and their families. This is often the only option for single women living alone on a single person's rations in a camp or paying rent in an urban area. Refugee women and girls are among the most marginalised people in the world and this marginalisation culminates in a range of human rights violations and abuses. To date there has been little or no legal redress and many perpetrators continue to function with impunity (Collated findings from 33 years in the field, Pittaway and Bartolomei)

This is a strong statement compiled, from evidence shared by literally thousands of refugee women from across the world, over 33 years. The story is of resistance to the acceptance of this major human rights abuse and the fight to change it. It has involved collaboration with a huge network of dedicated people at national, regional and international levels. On the journey, we have worked with activists, service providers, policymakers, academics and politicians. Most importantly, we have worked with many fantastic refugee women and girls, who despite all they have endured still have hopes and dreams, who can laugh and plan for a better future. That is what keeps us involved.

It is a complex story with many sub-plots, which range from refugee sites to the United Nations (UN), from the homes of refugees in camps and slums to the homes of refugees resettled in developed nations. It is a tale of tragedy, power, strength, resilience and political machinations. We include anecdotes which illustrate the raw politics involved in our experiences, the challenges and disappointments as well as the successes. There are stories that will make you cry, shout out in anger and others so ridiculous you just laugh out loud.

Rape in Conflict Is Not a New Story

'Rape and pillage' as the reward of conquering heroes has been an accepted outcome of conflict and war for millennia. Given the subordinate position of women in society throughout history, rape in conflict has been seen as their fate and seldom discussed except as a sad but rather insignificant outcome of war. These crimes against women were often dismissed as private acts, as 'a ubiquitous feature of social relations' (Mahmud, 1996, p. 370). It was not until the 1970s that the phenomenon of rape in war became the subject of research and study, and the notion of the inevitability of rape was finally challenged. In 1975, Susan Brownmiller wrote one of the first and most extensive historical accounts of rape in conflict. In this, she charted the history of rape as a strategy of terror from biblical times to present-day conflicts. She provided detailed historical evidence, supported by later accounts, of the widespread rape of women during the First World War in Belgium and in the Second World War in China, in Russia, in Germany and across Asia (Chinkin, 1994).

This is not just a historical note. Still today, the majority of refugee women and girls suffer from rape and sexual violence. In 2017, a Senior United Nations High Commissioner for Refugees (UNHCR) staff member stated that 100% of the refugee women fleeing conflicts in boats had been raped and sexually abused on their journeys (Bartolomei, 2017). Women attempting to escape the horrors of war via smugglers are advised to carry condoms, as rape is inevitable (Elbagir, 2018). Rohingya women, women from the Congo, Syria, the Central African Republic and South Sudan report rape as part of their everyday experience. We are already hearing similar stories following the fall of Afghanistan and from Yemen in 2021 (HRW, 2021; Narain, 2021). Tragically, in 2022, we are now hearing similar horrendous stories from the Russian invasion of Ukraine.

How Our Journey Began

This story began for me in 1989, when I was employed by a recently established non-government organisation (NGO), the Australian National Consultative Committee on Refugee Women (ANCCORW), to conduct a research project with refugee women from 19 different countries resettled to Sydney, Australia. During consultations with over 300 women, the research uncovered major problems in settlement service provision, but, totally unexpectedly, it revealed the extent of endemic rape and horrendous sexual torture the women had endured as part of the refugee experience, and the lack of response to this, both in Australia and internationally. Yet when I presented the findings, there was huge resistance from government, policy-makers and service providers to acknowledge and respond to it. I was effectively told that my evidence was incorrect and not to include it in my report. I was angry and determined that the women would not be silenced and that their courage in speaking out for themselves and their communities would not only be heard, but would lead to change in law, policy and service provision.

Little did I imagine that this would be the start of a life-long journey that would take me to work within the United Nations system and with refugee women globally. I was joined by Linda Bartolomei in 1998, and as the story progressed, we collected testimony and evidence from literally thousands of women from over twenty countries. In every refugee camp and urban slum we visited, women were determined to share their stories on record, on film, and if given the opportunity, to testify at forums and conferences. Alongside them, we took their fight to the highest levels of power. We have lobbied and advocated at the UN in Vienna, New York, Geneva and Bangkok. We have worked with UNHCR, the Human Rights Council (UNHRC), the UN Population Fund (UNFPA), United Nations Development Fund (UNDP) and UN Women. We have collaborated with many governments, partnered with major international NGOs and advocated alongside small unfunded community services and refugee-led organisations. We have lobbied strongly for refugees to have the space to have their voices heard in international and national fora, and to participate in problem identification, decision-making and identifying solutions for themselves, their families and their communities. Throughout this story, we chart what has succeeded, as well as the ongoing challenges to the achievement of human rights for all refugee women and girls.

Case studies provide an insider view of how UN bodies function, and how politics influences all aspects of law and policy development. This is the understanding that we had to develop in order to lobby and advocate effectively. Our knowledge about how to achieve change in the UN system developed in each stage of our journey as we went back and forth from working in the field with refugee women in camps, to lobbying in the halls of power. We have built a comprehensive picture of the UN system, its strengths and weaknesses. Most importantly, we learned how to function within it.

Academics or Activists? Daring to Say 'Action Research'

Along the journey, Linda and I have faced questions from within 'the academy' as to whether we are academics or human rights activists. After a combined 50 years in the field, we are proud to claim that we are both. Our academic work has laid the groundwork for significant changes in international law, policy, program design and evaluation. It has contributed significantly to the acknowledgement of, and response to, the endemic rape and sexual abuse, experienced by refugee women and girls. At the same time, at each stage, our understanding, analysis and action have been shaped directly by the refugee women with whom we have worked. We are not 'traditional' academics and often work 'outside the box'.

Our unconventional approach is embodied in the action research methodology we have developed, known as *Reciprocal Research* (Pittaway & Bartolomei, 2009). A very important aspect is a strong focus on developing ethical ways of conducting research with very vulnerable populations, and the methods we use challenge conventional wisdom about how research should be done. Our research aims to achieve social change. It acknowledges that people who agree to participate in research projects are not research subjects, or objects to be examined under the often-distorted magnifying glass of traditional research methods, but are partners who willingly share their stories in the search for a truth. They are also entitled to expect something positive from participating in the process. *Reciprocal Research* involves gathering stories from participants while giving back training, mentoring and opportunities to advocate for themselves and their communities.

More than once, we have been questioned about the use of 'stories' as research data. It has been suggested that stories are something women tell children: anecdotal, unmeasurable, subjective and unsubstantiated; of much less value than cold, hard quantitative 'facts'. Our response is that the stories we have recorded are evidence: hard data, both qualitative and quantitative, which can be triangulated, rigorously counted and analysed. Documenting one story is a case study; recording and organising the data from 300 women telling similar stories from the same refugee site produces a triangulation of qualitative and quantitative data; compiling the stories of 10,000 refugee women from across the globe becomes irrefutable evidence of systemic and endemic human rights abuses (Pittaway & Bartolomei, 2018). Stories are the subjective narratives that humans use to describe their experiences and shape their understanding of what has happened to them. Refugee women use them to articulate the inexplicable, to validate their experiences and their reactions to the abuse they have suffered. To deny the stories shared by refugees is to deny their experiences.

Other questions that have been raised about our methodology are about our subjective bias as refugee advocates. Some academics question whether research can be valid if it is coloured by the experiences of the researcher. Our response is 'can it be valid if it is not?'. Throughout these chapters, we grapple with these questions and detail how we sought to answer them through our developing methodology and rigorous academic analysis. We acknowledge and discuss the impact of our own world views, and how these changed over time, based on our own experience as researchers. We have been traumatised as we documented first-hand accounts of rape, torture, harassment and murder. Often we did this while living in appalling conditions in refugee camps, similar to those of the refugees—but always with our passports, cash and plane tickets in our pockets. We would then fly to the rarefied atmosphere of the UN in Geneva or New York, don our business suits to report on our findings. We make no apologies for the influence of our experiences on our research. Without these experiences, we would not have been able to gain the trust of refugee women nor the insight into their lives and the systemic disadvantages they face.

The Importance of the Theoretical Framework

At times, within the hallowed halls of the Academy, there is a sentiment that applied theory is of lesser value than the philosophy or 'pure theory' which some of our colleagues produce. We, of course, challenge this assertion. In recent times, however, following the trajectory of university funding models, more marketable 'research impacts' have come into favour. We argue that both are equally valuable. In the same way that in some faiths, contemplative nuns pray in silence for those nuns who labour outside the convents and thus have less time for prayer, so the pure theorists amongst our academic colleagues provide us with the enlightenment with which to interpret the findings of our applied work. In return, we test their theories, at times finesse them, and together we become part of a circle of advancing knowledge which is at the heart of the academic endeavour. This is not to imply that we regard our colleagues as nuns and monks. That would be taking the analogy too far!

Over the years, we have developed a theoretical framework to guide our reciprocal research and we hope it will achieve three things: the first is to explain what has guided our thinking and actions over the years; the second is to encourage people to explore and further develop the theoretical model themselves; and the third is to show the importance of academic theory and a rigorous application of this to understanding and interpreting empirical evidence. At each stage of this story, we have analysed the data through the lens of theoretical literature current at that time and used this to signpost the next stages of the research as we continue our involvement in the field. In answer to the question of whether we are advocates or academics, we can honestly say that without both the solid academic framework, which guided each step of the process, and the advocacy alongside the refugee women, our work would not have been as successful as it has been.

Politics from the Start—Refugee Policy in Australia

A brief history of Australia's response to refugees provides a backdrop to our story. It has major consequences for refugee women and forms the skeleton on which the rest of the story hangs, so please bear with it. Australia has been involved in the international response to refugees since 1921. The early response was guarded and strongly influenced by

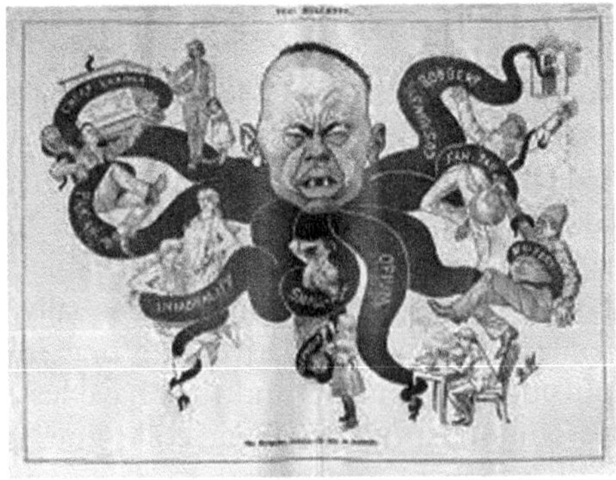

Fig. 1.1 The Mongolian Octopus—His Grip on Australia. Cartoon by Phil May, Published in The Bulletin, August 1886 (Image courtesy National Museum of Australia. Reprograher Lannon Harley)

the 'White Australia' policy,[1] which dominated Australia's immigration policy until 1975. When the UN established an International Refugee Organisation (IRO) in 1946, to deal with the European refugee crisis, Australia did not support the motion and refrained from voting, because it was feared that this could involve accepting refugees from non-Caucasian stock. However, due to increasing paranoia in Australia about invasion by 'the yellow hordes' from the north following the Second World War, the catch cry of 'populate or perish' was born and Australia began actively soliciting migrants. Since white migrants from the United Kingdom (UK) and its colonies could not be recruited fast enough to supply the demand, Australia's first Minister for Immigration, Arthur Caldwell 'became increasingly aware of the splendid human material to be found in the [European] refugee camps' (Markus, 1984). Importantly, of course, these refugees were white (Fig. 1.1).

In 1947, Australia joined the IRO and initially agreed to accept 12,000 people per year for a three-year period. Between 1947 and 1952, 181,700 refugee and displaced persons entered Australia through this scheme and related NGO sponsorships. The stringent health standards required for acceptance earned the country the international reputation of using refugees as 'grist for the labour mill'. The intake was seen to satisfy Australia's need for labour rather than its humanitarian obligations, and the government avoided responsibility for 'hard to place' refugees.

[1] The White Australia policy, which began in 1901, is the name given to a set of racist and discriminatory policies that prevented people of non-European ethnic origin (non-Caucasians) migrating to Australia. The policies were dismantled in the mid-1970s.

Women were seldom resettled in their own right, but as 'breeding stock' and a support to their worker husbands (Fig. 1.2). However, while the Australian Government gained a reputation for being intransigent in its attitude towards refugees, in contrast, Australian NGOs were recognised for a high level of involvement in the issue, both at home and on the international stage. This included lobbying and advocacy on behalf of refugees and the establishment of overseas aid programs and resettlement programs in Australia (National Population Council, 1991).

In 1950, UNHCR was established, and in 1951, the Refugee Convention (UNHCR, 1951) was adopted. It was designed to uphold the right to seek refuge and claim asylum for those who had fled persecution during the Second World War (WW2). In 1967, in response to the waves of refugees which were generated from successive wars and conflicts across the globe, the 1967 Protocol to the Refugee Convention (UNHCR, 1967) was adopted and signed by Australia. The Convention names three

Fig. 1.2 A woman's place is in the home (Australian Government 1954)

'durable solutions' for refugees, which are: local integration, usually in the country of first refuge; return to the country of origin, if safe to do so; and resettlement to a third country, such as Australia. Resettlement was an option which was initially only offered by a small number of developed nations, with numbers restricted by governments and a rigorous interview process for candidates to determine their ability to assimilate.

At first, Australia had a generous resettlement policy and from 1953 to 1997 accepted 199,000 refugees from across the world. During this period, by far the largest intake of refugees to Australia was from Vietnam. More than one million Vietnamese people had fled persecution, mainly to Thailand, Hong Kong and Malaysia, escaping the war involving the United States (USA), Vietnam and Cambodia. More than 2000 made it to Australia by boat, beginning a national backlash against 'boat people' which continues to this day. Sadly, this also spread to other developed nations. In 1975, an Interdepartmental Committee on Refugees was established by the Australian Government and NGOs were invited to participate on that committee as recognised contributors to refugee resettlement. I was appointed as the representative of 'The Smith Family', a charitable organisation which was providing services to refugees.

1979 saw the negotiation of an Orderly Departure Program (ODP) between UNHCR and, initially, the Socialist Republic of Vietnam to promote the departure of refugees from Vietnam to countries of resettlement without the necessity of fleeing by boat. Australia accepted a large number, with smaller intakes from Cambodia and Laos, and from conflicts in several countries in Latin America with whom Australia had politically strategic ties. Under pressure from UNHCR, Australia also began to accept refugees from African nations. While these arrivals met Australia's humanitarian obligations, they were also essential to the demands of a burgeoning labour market, initially providing mainly unskilled labour.

Australia introduced a system of refugee selection in 1979. Government officials were urged to select those refugees most likely to resettle successfully in Australia. These were the young, the healthy, the well-educated and people with a family support system already established here. This policy actively prevented selection of those with an obvious disability or those likely to require substantial social support. The many refugees who did not fit the stringent requirements of developed countries remained in the camps, often for years. This included many widowed and single women living with a high risk of violence. The problems which had led to them not being accepted in the first place were usually exacerbated by the physically and mentally debilitating conditions of camp life (National Population Council, 1991).

At first, when refugees were resettled to Australia, they were offered the same settlement services as the waves of migrants then encouraged by the Government to fill labour needs after the Second World War. Most services were provided by the NGOs and increasingly by community services established and run by the migrants themselves. By 1985, it was obvious that the programs available to refugees were failing to address their needs. Once again, NGOs lobbied the Government for an increase in services, specifically recognising the need for specialist and tailored services for this group. In 1975, the Government established an inter-sectorial committee to coordinate the provision of material aid and on-arrival assistance to refugees entering Australia through settlement services. This was followed by a number of key reports, such as the *Galbally Report* (1978). This report made recommendations about the need for Australia to come to terms with the changing nature of Australian society brought about by immigration policy and adopt the concept of multiculturalism, which recognised the need to respect the culture of refugees and immigrants, rather than focusing on assimilation.

In the mid-1980s, a worldwide recession and the advancement of technology contributed significantly to unemployment in Australia. Debate raged about the level of migration needed to sustain economic growth and health, and the refugee issue was embroiled in that argument. Some politicians, and many in the Australian community, demanded that the migrant and refugee intake be cut. Others went as far as demanding recent arrivals be sent back to where they had come from on a 'last arrived, first to go' basis, regardless of the situations they had fled. Far from being a humanitarian imperative, this public position only highlighted the political nature of the refugee response.

In the face of this political maelstrom, the NGO sector continued lobbying for the rights of refugees. In an attempt to address the issue, the Committee to Advise on Australia's Immigration Policies (CAAIP) was established and produced the *Fitzgerald Report* (1988). While recommending several positive changes in the provision of settlement services, it stated that they should last no longer than two years, by which time refugees should be able to access mainstream services. The community sector argued strongly against these findings as their experience indicated that settlement is a lifetime experience with different needs emerging at different times. Disagreements on these policies not only arose between political parties, but also between the NGO and community sector players, who also came from a wide range of ideological positions.

A Grant in Aid (GIA) program was introduced to employ people, preferably bilingual and from the new communities, as community development workers, with the aim of promoting multiculturalism and the more effective integration of refugees and migrants into the [white] Australian community. Following the UN Decade for Women, 1975–1985, and the UN Nairobi Women's Conference 1985, world attention began to focus on the plight of refugee women. In response to commitments made at the Nairobi conference, and at the request of UNHCR, in 1988, the Government established a Women at Risk (WaR) Visa category within its migration program, with an annual allocation of 60 places. It provided opportunities for resettlement in Australia for refugee women and their dependants, without traditional male protection, who had either experienced violence, including rape, sexual and gender-based violence, or were in vulnerable situations. It was anticipated that the 60 cases would be accepted in the program in 1989 and 1990 with the possibility of expanding that number in the following year. The Government also at this time established the Bureau of Immigration Research (BIR), with significant funds. Their role was to support university and community-based research into the experiences and needs of migrants and refugees in Australia. While these apparently generous responses to refugees must be seen as welcome, the Australian Government used their generosity in this area of the refugee resettlement response as a counterpoint to its increasing refusal to accept that asylum seekers had the right to travel to Australia and claim refuge. It was against this backdrop that I joined ANCCORW to undertake research into the settlement needs of refugee women.

A Complex Story

While the book charts a process of achievement, it is not a straightforward, linear, well-mapped journey. There are dead ends, roads which circle back to where we were or even further behind, landslides, huge potholes and quite a few roads with incorrect signposts! We have put indicative dates at the beginning of each chapter to assist you to navigate these twists and turns. However, we have managed to navigate the tortuous route and are pleased with the achievements we have made along the way. The process has been at times painful and frustrating, but overall, it has been a privilege because of the amazing refugee women, activists and human rights champions with whom we have worked over the years. While we have attempted to present this journey in a logical manner, at times this has just not been possible, but we have done our best to make it as clear as we could! We hope this 'map' will prove a useful guide.

References

Bartolomei, L. A., Dixson, T., Sheikh-Eldin, M., & Zahau, C. (2017). *Gender audit report on the second and third thematic discussions on the global compact on refugees*. Forced Migration Research Network UNSW, Sydney. Retrieved November 16, 2021, from http://www.unhcr.org/en-au/events/conferences/5a251b537/gender-audit-report-second-third-thematic-discussions-global-compact-refugees.html

Chinkin, C. (1994). Symposium—The Yugoslav crisis: New international law issues—Rape and sexual abuse of women in international law. *European Journal of International Law, 5*(3), 1–17.

Elbagir, E. (2018, February 27). *Human trafficking in Nigeria: 'Don't struggle if you're raped'*. CNN. https://edition.cnn.com/2018/02/27/africa/nigeria-migrant-smugglers-intl/index.html

Fitzgerald, S. (1988). *Report of the Committee to Advise on Australia's Immigration Policies*. Parliamentary Paper (Australia. Parliament), 335. Retrieved November 16, 2021, from https://catalogue.nla.gov.au/Record/1920991

Galbally, F. (1978). *Review of post arrival programs and services for migrants*. Migrant Services and Programs Canberra. Retrieved October 20, 2021, from http://www.multiculturalaustralia.edu.au/doc/galbally_1.pdf

Human Rights Watch (HRW). (2021). *World report 2021: Yemen*. Human Rights Watch. Retrieved November 16, 2021, from https://www.hrw.org/world-report/2021/country-chapters/yemen#351320

Mahmud, N. (1996). Crimes against honour: Women in international refugee law. *Journal of Refugee Studies, 9*(4), 367–382.

Markus. (1984). Cited in National Population Council (NPC). (1988). *A proposed settlement strategy*. AGPS.

MasterClass. (2021, September 1). *Writing 101: How to write a prologue*. MasterClass. Retrieved November 16, 2021, from https://www.masterclass.com/articles/writing-101-what-is-a-prologue

Narain, V. (2021, August 16). *The world must not look away as the Taliban sexually enslaves women and girls*. ABC News. Retrieved November 10, 2021, from https://www.abc.net.au/news/2021-08-16/afghanistan-taliban-women-and-girls/100379914

National Population Council. (1991). *The National Population Council's refugee review July 1991*. AGPS.

Pittaway, E., & Bartolomei, L. (2009). Innovations in research with refugee communities. *Transitions, 21*. Retrieved September 10, 2021, from https://startts.org.au/media/Refugee-Transitions/Refugee-Transitions-Issue-21-innovations-in-research-with-refugees.pdf

Pittaway, E., & Bartolomei, L. (2018). *From rhetoric to reality: Achieving gender equality for refugee women and girls*. World Refugee Council Research Paper Series 3. Centre for International Governance Innovation, Ottawa, ON. Retrieved November 10, 2021, from https://www.cigionline.org/publications/rhetoric-reality-achieving-gender-equality-refugee-women-and-girls

United Nations. (1951). *Convention relating to the status of refugees*. United Nations Office of the High Commissioner for Human Rights, Geneva. Retrieved November 16, 2021, from https://www.ohchr.org/EN/ProfessionalInterest/Pages/StatusOfRefugees.aspx

United Nation. (1967). *Protocol relating to the status of refugees*. United Nations Office of the High Commissioner for Human Rights, Geneva. Retrieved November 16, 2021, from https://www.ohchr.org/EN/ProfessionalInterest/Pages/ProtocolStatusOfRefugees.aspx

CHAPTER 2

Opening Pandora's Box: Rape, Sexual Abuse and Refugee Women: 1989–1990

Narrated by Eileen Pittaway

> *We have tried to tell people, but no-one will listen. They don't want to hear. They say women will not talk about rape because we feel ashamed. Who should be ashamed? Us, or those who raped us?* (El Salvadorean refugee woman 1990)

In 1985, the United Nations held the Nairobi Third World Conference on Women with the goal of achieving gender equality for women everywhere. One of the key areas of concern was refugee women and girls, named as one of the most vulnerable groups in the world. As well as problems in accessing food, safe shelter, health and education, rape and sexual abuse were clearly identified. Commitments were made by all governments, including Australia, who signed the Outcomes Document, to improve the protection of refugee women and girls worldwide. At the end of the meeting, the UN declared a 'Decade for Women', and governments were urged to act to address gender inequality. An 'International Working Group on Refugee Women' was formed and, in 1988, convened a meeting in Geneva to develop strategies to follow up commitments made in Nairobi. While these included addressing rape and sexual abuse, the estimates of the proportion of women who were victims were extremely low—only around 6–7% (Kelley, 1989). A group of women from Australia, including a refugee activist named Velvy Holden, a refugee from Colombia named Maggie Escartin and a refugee from Armenia

E. Pittaway and L. A. Bartolomei,
Only Rape! Human Rights and Gender Equality for Refugee Women,
Sustainable Development Goals Series,
https://doi.org/10.1007/978-981-19-0916-0_2

named Hermine Partamian, attended the meeting. On their return, they formed a small unfunded organisation, the Australian National Consultative Committee on Refugee Women (ANCCORW), to advocate with, and on behalf of refugee women, with the specific aim of lobbying the Australian Government to fulfil these commitments.

Without these amazing women, this story would not have evolved in the way that it has. They decided that an appropriate strategy for local action would be to conduct research into the resettlement needs of refugee women in Australia, and an evaluation of how well these needs were being met by settlement service provision. They applied to the Federal Government Bureau of Immigration Research (BIR) to fund the research and to the Department of Immigration, Local Government and Ethnic Affairs (DILGEA) for a Grant in Aid (GIA) worker. To their astonishment, they received both. The Refugee Council of Australia (RCOA) offered to auspice ANCCORW and host them in their offices, but, at first, there was very little interaction between the two organisations.

Joining the ANCCORW Team

I was engaged to conduct the research project for ANCCORW, with academic supervision from Dr Barbara Ferguson at the University of New South Wales (UNSW). It is relevant to the story to include my history at that time. Prior to migrating to Australia, I had worked in international development in both India and Peru, and spoke functional Spanish. Subsequently, I had worked with resettled refugees in Australia for over ten years with The Smith Family, a large Sydney-based charity. During this time, I interacted with refugee women resettled to Australia and had been invited onto the boards of several refugee-led organisations. I also represented The Smith Family on a number of state and national government committees on refugee resettlement. I left that position following the birth of my daughter and worked as a teacher at Technical and Further Education (TAFE) in the School of Welfare and Community Development. In this position, I developed and delivered courses for migrants and refugees to supplement and assist recognition of their previous social and community work qualifications, and to provide them with qualifications to work in the emerging employment sectors created by the Committee to Advise on Australia's Immigration Policies (CAAIP) (Fitzgerald, 1988) and the GIA scheme.

I also worked as a researcher with people with a disability, with a colleague, Marie Wilkinson. We used action research in the projects we undertook and actively included participants as research assistants. This included providing training and support and ensuring that at least some of the funding flowed into the pockets of the communities concerned. For example, when researching the needs of people with a disability, we recruited research assistants from each of the disability groups, including people with an intellectual disability. This was a controversial move, dismissed by some to be 'unscientific' and not academically rigorous. However, it produced excellent results, and many of the recommendations from the community were implemented by government agencies. As researchers, we were a vector for the participants, rather than the authors of the findings.

With this as my background, I was really excited by the BIR research opportunity. I had built extensive networks in refugee communities and a strong appreciation of what they were experiencing in Australia. I found the thought of conducting the work as an action research project exciting, calling on my many contacts to work with me as bilingual research assistants and interpreters. The ANCCOW selection panel was impressed with my previous experience and established networks, the fact that I would be able to speak directly with the refugees coming from Spanish-speaking countries, and the proposed action research methodology. To my delight, I was awarded the position. The salary offered was considerably less than I was earning, so it was agreed that it be a four-day per week position, giving me freedom to undertake other research. The convener of ANCCORW, Velvy Holden, was pivotal in the project, working closely with me in the research and the advocacy.

Unfortunately, there was some criticism from the academic sector that the position was not awarded to a university-based academic with a doctorate, as was the more usual practice for large government research grants. This provoked me to do what I had been considering for a while, to put my extra day a week to good use and enrol in a PhD. This course of study, which I completed over the next eight years, provided the opportunity to thoroughly explore and develop the rigorous academic framework used throughout the story. It taught me the importance of theoretical research to explain and support actions, and, the most important lesson, that successful advocacy and academic work are not mutually exclusive. In fact, they are intertwined.

When I was awarded the position, the BIR was somewhat taken aback by the fact that I would not be employing a more traditional research methodology. They were keen on a quantitative methodology to provide solid numerical evidence of what was happening, but since it was community, rather than university based, they were persuaded to see how the action research developed.

The Research Project

The stated aim of the project was to examine the resettlement needs of refugee women in Australia and to evaluate if they were being addressed by the current service provision. The ANCCORW Board also wanted to know the experience of women who had come to Australia as part of the new WaR visa category. The research was structured to encourage participation by refugee women in identifying their own needs and suggestions for solutions. The project management committee, comprising members of the ANCCORW Board and UNSW staff, had identified that, at many levels, the language addressing refugee women needed to be challenged. They believed that emphasising the notion of 'protection' of refugees, which is inherent in the 1951 Convention and 1967 Protocol, caused refugees, especially women, to be disempowered by the implication of vulnerability. It was hoped that active involvement in identifying solutions to their problems would contribute to the empowerment of refugee women. This was primarily accomplished through the employment of refugee women as interpreters and co-researchers. These women were bilingual workers who were known to me through the TAFE course they had completed and were respected in their communities.

When the research questions were being developed, the issue of exploring possible experiences of torture was discussed by the steering committee. The international literature survey and anecdotal evidence at that time had suggested that approximately 30% of refugees, mainly men, had suffered from torture and severe forms of trauma. There was some small mention of sexual abuse of women, but it was not considered central to the issue. It was suggested that women were more likely to suffer secondary trauma based on the experiences of their male relatives. To test this assumption, towards the end of the interviews, I asked the women a gentle question. 'We hear that refugee women sometimes suffer from violence as part of their experience. Do you know any refugee women this has happened to?'. We had no idea that with this single question we would be opening Pandora's Box!

The project commenced in March 1990, data collection began in April, and it was agreed that the management committee would review the findings at the half-way point of the interviews, to decide if any adjustments needed to be made. It was planned that I would interview 120 women from 19 different countries. Interviews were designed to last one hour, focus groups for two hours. Working closely with the interpreters, every effort was made to put the women at ease. The purpose of the interviews was carefully explained, and demographic data was not collected until the end of the process. The questions sought the opinions of women about the effectiveness of current service provision, from both the non-government sector and government services, and what they would like to see changed in future.

Comfortable because of the presence of a trusted bilingual worker, the women were remarkably frank in their assessment of the services offered, and the ways in which they were provided. While the overwhelming feeling was of gratitude to Australia because of the security they had found here, they also reported racism, discrimination, and a lack of respect from many service providers. They shared graphic and disturbing examples of major problems that were experienced in the provision of English classes, access to social services and a lack of employment services, orientation and information services, and secure, affordable accommodation. The NGO sector did not escape, with many women reporting feeling judged and shamed at having to ask for charity. The women also shared their feelings of loss of homeland and family and friends, and their grief about those who had died—over one-third of the women had lost at least one child in conflict or flight. They had excellent suggestions for ways in which the services could be made more effective; most importantly, these included drawing on the knowledge, skills and experience of the refugees themselves. They had a lot to offer and wanted to be contributors to their new country, not burdens.

At a focus group held at the request of 10 women from Vietnam, there was a huddled conversation in Vietnamese during the tea break, after which one woman disappeared. She came back, triumphant, with a lottery ticket, which she presented to me saying, 'This is from all of us – if it wins, we will set up all the services we have been discussing'. Unfortunately, we did not win the money, but the gift was a sign of the determination of those women to fight for their rights and gave me the inspiration to continue fighting with them.

Unintended Consequence and Unexpected Findings

When we got to the question about the violence they had experienced as refugees, it was like a dam wall breaking. Suddenly, stories of rape, sexual abuse and sexual torture flooded out. Interviews extended into hours, and I was often asked to return at another time, which I did. Focus groups were extended and I was asked to hold second groups with the same women, to continue the conversation. As the stories came pouring out, the women became more comfortable sharing their experiences. Frequently, they would start responding in the third person, telling me about 'my cousin', 'my friend' or 'my neighbour', but this often changed to the first person as the story was told. For example:

'my cousin was walking home from the well…' and then 'I was attacked and raped by three men. They tore my clothes and beat me. I was left bleeding on the path'.

Over and over, these stories of rape and sexual abuse were told with deep and graphic detail. Over and over, women said that no one had ever before given them permission to talk about them. If they were asked directly if they had been raped, they said 'no', fearing it would impact their chances of resettlement or be used to shame them. More often, it had not ever been asked, and if they had obliquely tried to bring it into conversations, it had been ignored. As one woman said, 'sometimes we try to open the door, but no one comes in'. Many tears were shed, but the women thanked me for the opportunity and said they were so relieved to be able to talk.

Juana's story, 1990: A failure of settlement service provision
Juana is 28 years old and has come here as a refugee from Central America. She is a widow with a 3-year-old daughter who has been in Australia for six weeks. She is staying in a government flat which she has been allocated for a period of 13 weeks, after which she will have to find accommodation on the private rental market. She speaks no English. Before coming to Australia, Juana had been imprisoned for two years because of the political activities of her husband. He was killed while she was in jail. While she was imprisoned, she was repeatedly raped, and suffered from other forms of physical torture - she is still receiving medical treatment for some of the wounds. Her parents took her one-year-old daughter and raised her as their own, in order to protect her from the police (family harassment is commonly used by authorities in Central America). Eight weeks before she came to Australia, Amnesty International, working with a church group, managed to get her out of prison and arranged for her and her daughter to be resettled. Upon arrival in Australia, she was given a temporary government apartment and a quick orientation to services in Australia. Fortunately, there are a few Spanish speaking people nearby who helped her, and two part time resettlement officers who offer assistance. Otherwise, she is almost completely on her own. Every day, her daughter screams for her 'mother' (really her grandmother). Every night, Juana screams in fear of the memories of her ordeal, and the terrible loneliness she is suffering here. She is terrified about having to find housing herself. She is receiving help from a torture and trauma counselling service, and a doctor has given her sleeping pills, but she says, 'I am too scared to take them - what if anything happens to my daughter when I sleep?'. The neighbours have complained to the New South Wales Department of Family and Community Services requesting that because the child screamed so much, she should be taken away from her (Pittaway, 1991). Appropriate assistance, including counselling, was finally provided for Juana and her daughter. She went on to study and qualified as a community worker. She now works to assist newly arrived refugees.

By late June 1990, upon completion of the first 70 interviews, the review was held with the management committee. Three distinct themes were emerging from the data. The most disturbing was that the extent of torture and trauma experienced by refugee women resettling in Australia was substantially higher than previously believed, and the majority of it was of a sexual nature. The sample of the first 70 women interviewed suggested that at least 80% of them had experienced either rape or systemised sexual abuse as part of their refugee experience, and this was a form of torture.

The second, which directly addressed the aims and objectives of the funded research, was that resettlement services and the refugee intake process were not adequately addressing the needs of refugee women. They were more complex than those of migrants, and there were problems in service delivery. The extent of loss and grief, torture and trauma reported, made the situation far more complex than previously thought.

The third significant finding was evidence of problems linked to attitudes about rape and sexual abuse when processing refugee applications in Australian overseas posts. This was particularly concerning with regard to the Women at Risk (WaR) Program. Most of the women were not even aware of the Program, and despite the fact that they all knew of women who had been raped and abused, and who were still living in extreme danger in camps and urban settlements, the very low annual quota for the WaR Program had not been filled in 1989 or 1990. It was becoming increasingly apparent that the WaR Program was failing to achieve its objectives because of the lack of acknowledgement of systemised rape and abuse as torture and trauma, and the system was failing desperate refugee women.

Ly's story, 1990: A failure of the Women at Risk Program

Ly came originally from Cambodia but was in a Thai refugee camp for eight years. She has been in Australia for two years and lives in a western suburb of Sydney with two of her teenage children, her two-year-old daughter and five-year-old grandson. Two other children died in the refugee camp and two teenage daughters remain in the camp. Ly's husband was killed in front of the children as soldiers sprayed bullets into their hut, and Ly was then raped at gunpoint. One of the children has a bullet wound. Ly's elder daughter is a widow with a 5-year-old son; her husband was killed as he left the camp searching for fruit for her when she was pregnant. Since then, she has been severely traumatised, and is only marginally functional. Her mother has to do everything for her and her small son. This young woman has been assessed for an invalid pension and is awaiting the outcome. No treatment has been offered to her: she seldom leaves the flat, has no friends or social contact and does not attend any English classes. Ly is illiterate in her own language, and the children have had little formal education in Cambodia; they are having problems adapting to the education system in Australia. The family is living in desperate poverty, having had to repay a loan to family members who sponsored them here and paid their fares, their accommodation is overcrowded and poorly furnished. Despite these problems, they are glad to be in Australia and thankful to the government for giving them refuge. Ly's dearest wish is to be able to sponsor her two remaining daughters from the refugee camp (Pittaway, 1991). It is important to note that Ly herself could have been eligible for resettlement as a Woman at Risk. Her two daughters still in the camp were definitely at risk and yet it had been impossible for us to find a way to have them assessed for the program. Professional help was provided to Ly, but the outcomes for her and her family are unknown.

It is worth noting at this stage that my actions in responding to these cases were questioned. Was it the role of a researcher to take action, or merely record? It brought to mind the conundrum often used in philosophy classes. *Should a journalist stop filming a massacre and try and help?* It was a major ethical issue which has resonated throughout our work.

How the Findings Were Received Mid-Point

Because the initial findings were unexpected and potentially so explosive, my methodology immediately came under scrutiny. In particular, the torture and trauma findings, which were so much higher than those identified in the initial literature review, were rigorously examined by the academics on the ANCCORW committee, and my method for selecting the research participants was questioned. When no problems were identified, the sample size was increased to further test the findings. I eventually formally interviewed 204 women and conducted focus groups with more than 100 others who approached the bilingual workers because they wanted to share their stories. Several of the bilingual workers also requested to be interviewed. The flood of stories of rape and sexual abuse continued, and they came from all the 19 countries covered in the research—they were the stories of refugee women everywhere. To organise the data, I developed a classification for codifying torture and trauma for the project (Pittaway, 1991). This was later used by torture and trauma services in Australia.

It was a difficult time for me as the researcher. In total, I documented over 300 stories of rape, torture and sexual abuse. One traumatic realisation was that before doing the interviews, I had considered myself as something of an expert on refugee women. I had worked with them, talked with them and—I thought—built trust with them. How wrong I was. In fact, I had been one of those people they described; I had been someone who closed the open doors offered to me and did not give permission for them to speak out.

The sheer scale of the problem and the graphic detail provided by the women was also incredibly traumatic. I included some accounts in the final report, and the editor employed by BIR handed the job back because she found them too confronting. By the end of the research, I was suffering from secondary trauma (although this was neither recognised nor discussed back then), and for quite a few months, I found it difficult to relate to men at all except as potential rapists. It took the love and support of my family and circle of friends, including some fantastic men, to enable me to get things back into perspective. Yet the pure volume of what we had uncovered made me realise that the abuse was not just perpetrated by a few bad men. Rape and sexual abuse were a usual part of a refugee woman's experience and were perpetrated with impunity by men who saw it as their right. It was a chilling realisation about the way that the rape and sexual abuse of refugee women were normalised and tolerated.

I was also taken aback and traumatised by the fact that many people in the field refused to accept the research findings. At that stage I was very naïve. When I first documented the extent of the rape and sexual abuse experienced by refugee women and girls, I honestly believed that once this data was presented to policy-makers and service providers, they would be as equally shocked and disturbed as I was, and that actions would immediately be taken to address the issues. *This did not happen.* As the findings were revealed, there was significant pushback, both from government service providers and from major charitable organisations which had been criticised by the refugee women. In fact, I was told I was exaggerating the numbers, 'making it up', that I was a 'man hater', a crazy feminist. Other people told me that even if the women's stories were true, it was best not to openly address it, since it was not in the women's culture to discuss rape, and it would shame not only the women involved, but their whole communities. I asked myself: 'If this was true, why had over 300 women resettled to Australia from 19 different countries shared their stories of rape and sexual abuse? Why had they stated that this had to stop?'.

We also received pushback about the finding that the Federal Government's WaR Program was failing. We were not allowed to include this in the research report, as, it was argued, it was not in the research specifications. We were told unofficially that the Government had insisted it not be included. Under some pressure from BIR, the ANCCORW Board decided to publish only the findings about the resettlement experiences of refugee women in the research report and to pursue the findings about the failure of the WaR Program, separately. As is explored below, once launched, the report, *Refugee Women: Still at Risk in Australia* (Pittaway, 1991), went on to have a life of its own. It raised awareness of the experiences of refugee women and girls prior to arrival in Australia and had a significant and positive influence on the provision of services to refugee women and girls across Australia and internationally.

The research report detailed the horror of refugee women's experiences, but also highlighted their strength. Despite their unbearable stories, the women were incredibly resilient. They were determined to make a success of their new lives in Australia, and resolute that their children would succeed. They had many suggestions about ways in which better targeted service provision could help them to achieve these goals, including the provision of culturally appropriate psychosocial support services. Their recommendations became the focus of ANCCORW's advocacy with the Australian Federal and State Governments, as well as informing our international work. There was already a network of strong refugee women leaders in Australia. Some of these joined ANCCORW and groups were formed in several communities to provide mutual support and to develop lobbying strategies. Increasingly, the women spoke out on their own behalf.

Validating Action Research

At the beginning of the project, the action research methodology was considered to be too overtly political by the Director of the BIR, Professor John Nieuwenhuisen AM. I fought hard, and he suggested that while maintaining its participatory nature, the research be referred to as 'social research'. However, over the course of the project and before the final report was published, he became convinced of the value of the action research and the validity of the outcomes. Towards the end of the project, he arranged for me to conduct a half-day workshop for the staff of the Bureau. This was daunting, as they were all male, PhDs and professors;

however, it went down well. When later discussing the international launch of the report, *Refugee Women: Still at Risk in Australia*, I requested that we have something other than the very plain black and white standard cover of the BIR reports. I was told firmly that their reports had to look uniform. I had not seen the final version until the actual launch, where with a large grin, John produced it at the ceremony. Sporting the white, green and purple colours of the feminist movement, it also contained a red streak woven between them, signifying, as he said, 'the suffering of the refugee women'. He described the work as 'ground-breaking' and the report became the most reprinted and cited report ever produced by BIR. He, one of my most important mentors, gave me the confidence to always push back when challenged about my work.

Theory and Discussion

While the research findings produced some very positive outcomes in settlement service provision in Australia, we were left with a number of unanswered questions. The first was why the extent of rape and sexual abuse suffered by refugee women was not recognised and addressed in policy designed to assist their settlement in Australia. The next was why, given the scale of abuse suffered by women in camps and urban refugee sites, the WaR Program was failing to fulfil its very small quota, and finally, why were so many roadblocks being put up to prevent me from exploring this further.

I now introduce the analytic tools and academic theories I used to help me understand and move forward. They are predicated on the assumption that the problem under the microscope is one of policy failure, building on work started in my master degree in social policy and continued throughout the story. It involves using three very different tools of policy analysis: the rational, the political and the critical paradigms, to examine policy formulation, implementation and outcomes. While it has not always provided answers to the inexplicable, it has guided me about where to look next. I mainly applied the rational and political paradigms to analyse this first stage of the research.

The Rationalist Paradigm

> *The critical distinguishing feature of organisation theories taking the rationalist perspective is the element of conscious, foresightful action, reasonably autonomously constructed to achieve some goal or value.* (Pfeffer, 1982, p. 7)

I first consulted the rational paradigm for answers to the conundrum I was facing, and very quickly abandoned it. The early writers in this paradigm of policy analysis presented a straightforward view of the function of policy (Lasswell & Kaplan, 1950; Simon, 1957). They saw it as formulated to address an identified issue, often a social problem or some aspect of social life over which government has jurisdiction and formal control. It usually involves the provision of a service or program to address the issue. A document is produced that describes the issues, the activity and its goal. The 'policy' is then implemented and evaluated. This view, in its purest form, is referred to in the literature as comprehensive rationality. It is predicated on a belief that 'policy elites', usually politicians or bureaucrats, have the formal power and authority to design and implement policy, and provides a model whereby empirically proven facts about the social world are used in policy formulation to provide answers to problems. The core assumption is that policy-making is a rational process of decision-making; a logical, hierarchical model of policy analysis based in psychological behaviourism to explain how policy is formulated, implemented and evaluated. It seeks to remove the 'politics' from government; indeed, to remove politics from the whole policy process altogether.

I had already begun to seriously question this model, as my experience as a policy analyst showed it to be a false premise. It did, however, enable insights into the way in which other protagonists in this story were functioning. I must admit that both I and the ANCCORW management committee had dearly wanted a rational approach to work. If this had been the case, how simple our task would have been! As the issue of rape and sexual torture was now identified, it should have been immediately accepted, new policies written to address settlement problems, and specialist torture and trauma services put in place for refugee women and girls. Steps would also have been taken to ensure that the WaR Program functioned smoothly. However, while some of this did eventually happen, at first, every attempt was made to discredit our findings both in Australia and internationally.

The Political Paradigm

> *Public policy analysis involves observing politics in action, tracing how economic and social forces, institutions, people, events and chance interact. It requires argument about cause and effect.* (Davis et al., 1988)

The work of academics or writers who accepted that the policy process is inherently political provided glimmers of hope. They acknowledge the contested nature of government policy and argue that, to make sense of it, we must understand the political nature, power play and dynamics of the process. They state that the policy process also has to be understood in its historical and social context, acknowledging the power structures, values and priorities of various actors who take part at any given time. The core assumption is that policy is not a constant, but the product of continuing interplay between groups. This means that our research had to be examined in the context of the history of Australia's refugee policy, which had always been more political than humanitarian. While at the time of the research project, the Government was taking some very positive steps forward with regard to refugees, it was in response to voter unrest due to a recession, and political contingency, both at home and internationally. We still needed a significant migrant intake to maintain economic growth, and we used our reputation as a generous country of resettlement to distract from our unwillingness to offer asylum.

In this model, problem definition is a critical part of the political process (Rochefort & Cobb, 1993; Weiss, 1989). The way the problem is defined legitimises some solutions over others. The complex dynamics between interest groups which contribute to the social construction of issues and the role of cultural values and ideology are acknowledged as part of the process. Instructions are often ambivalent, and decisions made 'on the ground' are built on a complex web of influence held by different players (Yanow, 1987). This process results in policy that is non-rational; some policy is not implementable, and sometimes it is totally 'un-doable'. Sometimes the policy process involves deliberate symbolic policymaking on the part of politicians and senior bureaucrats. Many policies are framed without reference to the underlying forces that are likely to undermine them. This was the challenge we had to face as the ANCCORW Board. We had changed the problem definition and questioned the policy framework currently in place. In doing so, we challenged the power structure, both within government and also the power of major non-government actors, such as the Church and large charities. In the mid-1990s, Considine (1994) and Hoffman (1995) stated that most policy research does not question the political context in which it takes place, but either neutralises or reifies the structure, and therefore reinforces the status quo. The ANCCORW Board had already come to this realisation, and we were determined not fall into this trap.

The 'Sacred' and the 'Profane'

In struggling to understand the many political paradoxes observed in the process in which I was initially involved, I drew on a theory which I used in my master's degree, developed by Pieter Degeling (1980).[1] He describes two different ways in which actors experience and articulate the policy process, the *sacred* and the *profane*. The *sacred* is the formal, officially sanctioned view of reality, a classic rational approach to policy. It is how those in power talk publicly about policy, while away from the public domain they may use very different language. The profane view of the policy process is what actually happens, and incorporates the politics, irrationality, ideological positions and power dynamics of the policy process, which are often not openly acknowledged, even by those who consciously use them. Many actors, in particular the policy elites, talk about the sacred version of reality as if it were the absolute truth while in reality their actions are framed by the profane. Actors who believe in the veracity of the sacred feel very uncomfortable with the profane concept, seeing it as manipulation (Figs. 2.1 and 2.2).

Degeling used the pyramid model of organisations as an example of the sacred view of the policy process. Power flows from the top down and access to the decision-makers occurs upwards through the various levels of the hierarchy. Thus, it is difficult for someone at the base of the pyramid, or an external actor, to have any power over decision-making. The decision-making process is contained within the constructs of the pyramid and is impermeable to external influence. The profane version of organisational structure places any particular actor, regardless of their position in the official hierarchy, in the centre of a circle or

[1] This work was presented as lecture notes, Master of Administration, UTS 1980. Despite an extensive literature search and requests to Degeling, I have been unable to locate anything further he has written on this model.

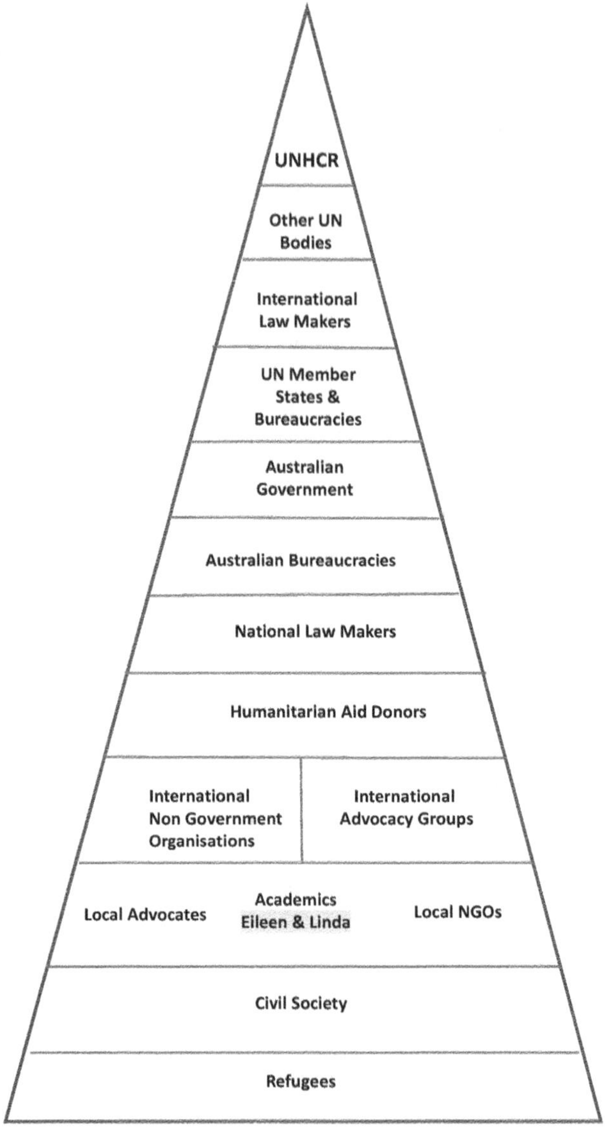

Fig. 2.1 Sacred view

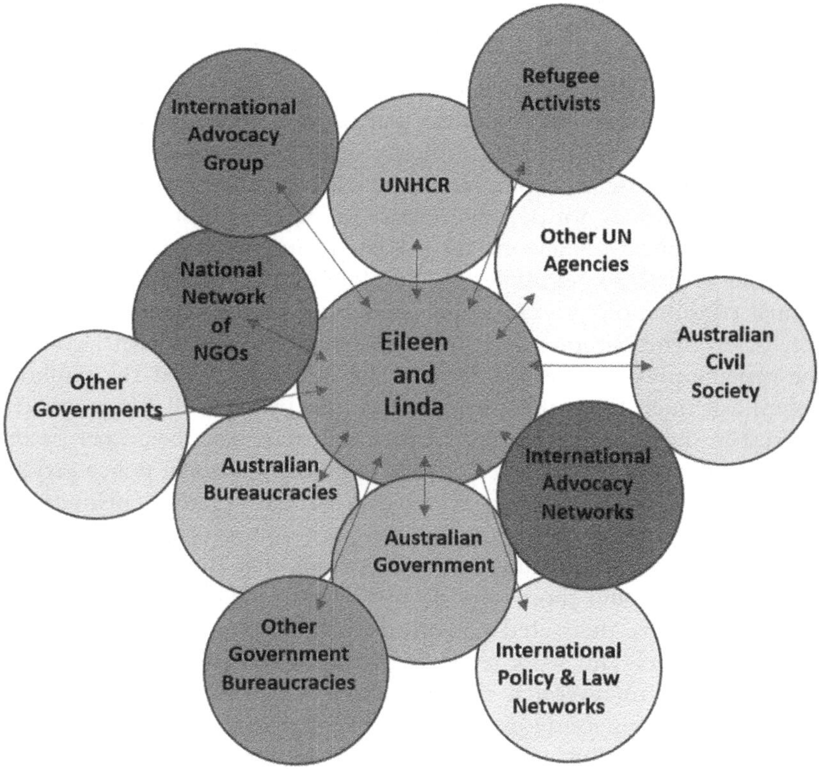

Fig. 2.2 Profane view

network of other actors with whom they interact. The better developed and stronger this network, the more likely an actor can influence decisions, often reaching decision-makers through informal channels which would be denied if they adhered to the sacred version of reality and worked solely within the prescribed framework. As shown in the diagrams, the sacred view puts myself and Linda very near the bottom of the power pyramid, but as we learned how to play the game of policy change, the profane view puts us right at the centre!

These concepts are manifested in several ways in the policy domain. Some actors completely accept and believe in the sacred. They see their role as implementing the sacred and controlling any actors who seek to question or subvert it. However, many actors such as politicians and bureaucrats, who apparently function in the rational framework, are merely presenting the sacred view of government and policy to the world. They are really extremely political players, and their game is profane. Often, they may be observed cynically using rationalist models as part of sacred discourse in their political activities as they seek to control a domain which does not pay due respect to the rationalist ideals. Insights from the political framework are co-opted into the rational framework to achieve this—for example, by developing and using tools such as formal consultation processes and appointing expert committees which are merely symbolic or tokenistic. It is policy-making as social control; the profane presented as the sacred. The intention is that the symbolic policy will quell the public outcry, and that people will forget or lose interest in the issue, believing that their demands have been met by the announcement of the policy. Thus, an apparently rational policy process is consciously used by policy actors in a covertly political manner. An important distinction can be made between those actors who identify as rationalist, but who are discursively conscious that they are part of the profane process, and those who do not have this level of consciousness. The latter believe the sacred version of reality despite activities which are actually profane. This work resonated with me so strongly I have incorporated it into my everyday thinking about the world we live in. I have continued to consciously develop an understanding of the profane and use this understanding in my analysis and my relationships with power brokers. Some may say this is a cynical approach, but it is one that is continually affirmed in my advocacy work, as you will see.

> In summary, the rational is how things should work, the sacred is how people say they work, and the profane is how they actually work!

The Dialectic of Change

As the ultimate aim of the research project was to achieve change, a useful set of insights came from Change Theory, seated in organisational management studies. This fitted firmly into the Rationalist Paradigm. The seminal work of Lewin (1946) spawned numerous reiterations of rational 'sacred' cycles of actions designed to bring about organisational change in a planned and orderly fashion. These did include the notion of radical change, when the systems were too broken to be mended, and had to be replaced, but this still fitted well into a rational paradigm. More interesting to me was the notion of the dialectics of change, a political and critical analysis which examined the oppositional positions of who decided change was needed, what did they want changed, and how could this be achieved? What opposition was there to the proposed changes, from whom and why? Finally, I liked the notion that change happens in at least two different ways, one being a staged progression, acknowledging the work of Lewin and others, but integrating the understanding of the dialectics involved. The second was a notion of catastrophic change, borrowed from the world of mathematics, when a totally unexpected and unpredicted element totally negates the change models used to date, and replaces them with a new paradigm based on the unanticipated reaction to the catastrophe (Phillips, 2000). Much of what ANCCORW achieved in the Australian context fitted the rational model. However, in a small way, the uncovering of the extent of sexual abuse experienced by the refugee women resettled to Australia triggered a 'catastrophic change'. It challenged many accepted wisdoms and current policy provision. However, the most significant use of this theory came much later.

Action Research—More Than a Research Method

You may wonder why I was so determined only to use an action research methodology when it is so contested. It is because I believe that it is the only truly ethical model of social research with vulnerable populations. Action research is a method that involves working *with* the research participants both to address their issues and to identify the research outcomes. A key outcome of action research is social change. The nature and structure of the approach engender a level of trust, and the invitation to not only discuss problems, but also to suggest solutions, shows a level of respect. I am sure that if I had used a more structured research methodology, such as a list of quantitative questions, I would not have collected the information which the women volunteered.

Action research is sometimes criticised for being too subjective and therefore not empirically valid (Macpherson et al., 2000). It is said to be a methodology for activists and not for academics. However, Fraser (1989) argues that there is in any case an artificial dichotomy between the two roles, that in fact activism and academia should be linked. Action research is a form of self-reflective enquiry, undertaken by participants in social situations in order to improve the rationality of their own social practices, their understanding of these practices and the situations in which the practices are carried out. The researcher must demonstrate self-knowledge and the capacity to 'learn to learn', including learning about their own part in the research process (van Beinum, 1998). Lewin (1946) defined it as participatory in character, democratic in impulse and simultaneously contributing to social science and social change. Action researchers aim to influence policy agendas to achieve social change (Gatenby & Humphries, 2000). The critical nature of the undertaking is 'not concerned with considerations of scientific objectivity or with the neutrality of the researcher ... it is interested in participative 'reconstructions' of social reality' (Macpherson et al., 2000, p. 51). This is exactly what we wanted to achieve with our work, and we think that we have succeeded.

The story of our research and advocacy undertaken over the last 30 years proceeds through a spiral of planning, acting, observing and reflecting. It involves many actors and key stakeholders who have an interest in maintaining collaborative control of the process (Carr & Kemmis, 1983). There are cycles within cycles: the cycles in the ANCCORW research process included undertaking the first 70 interviews, analysing the unexpected findings and then planning the next stage of the research. The next cycle involved processing the partial rejection of the findings about settlement services, the denial of the findings about sexual violence and the failure of the WaR Program. This spawned two new cycles of activity, one focused in Australia and one at the international level. The cycles undertaken over the years are reflected in each stage of the story.

Conclusion

The theoretical analysis I was using showed clearly that the rational paradigm of policy analysis was irrelevant to explain what was happening, except to understand the political posturing of the main actors. The overtly political nature of the process was obvious, but, at that stage, I had little understanding of why and what to do next. An opportunity was unexpectedly given to present our findings at the UN, and this opened a door which we had not previously even known existed. It led to our next cycle in the action research, which was planned with the ANCCORW

Board based on our findings and our analysis to date. Things were about to get a lot more contentious!

References

Carr, W., & Kemmis, S. (1983). *Becoming critical, knowing through action and research*. Deakin University Press.

Considine, M. (1994). *Public policy: A critical approach*. Macmillan.

Davis, G., Wanna, J., Warhurst, J., & Weller, P. (1988). *Public policy in Australia*. Allen & Unwin.

Degeling, P. (1980). *Organisational arenas and the practice of management*. Unpublished lecture notes, Kuringai College of Advanced Education.

Fitzgerald, S. (1988). Report of the committee to advise on Australia's immigration policies. *Parliamentary Paper (Australia Parliament)*, 335. Retrieved November 16, 2021, from https://catalogue.nla.gov.au/Record/1920991

Fraser, N. (1989). *Unruly practices: Power, discourse and gender in contemporary social theory*. University of Minnesota Press.

Gatenby, B., & Humphries, M. (2000). Feminist participatory action research: Methodological and ethical issues. *Women Studies International Forum, 23*(1), 89–105.

Hoffman, J. (1995). Implicit theories in policy discourse: An inquiry into the interpretations of reality in German technology policy. *Policy Sciences, 28*, 127–148.

Kelley, N. (1989). *Working with refugee women: A practical guide*. United Nations High Commissioner for Refugees.

Lasswell, H. D., & Kaplan, A. (1950). *Power and society: A framework for political inquiry*. Yale University Press.

Lewin, K. (1946). Action research and minority problems. *Journal of Social Issues, 2*(4), 34–46.

Macpherson, I., Brooker, R., & Ainsworth, P. (2000). Case study in the contemporary world of research: Using notions of purpose, place, process and product to develop some principles for practice. *International Journal of Social Research Methodology, 3*(1), 49–61.

Pfeffer, J. (1982). *Organizations and organization theory*. Pitman Publishing.

Phillips, T. (2000). *The catastrophe machine*. Retrieved November 10, 2021, from http://www.math.stonybrook.edu/~tony/whatsnew/column/catastrophe-0600/cusp1.html

Pittaway, E. (1991). *Refugee women: Still at risk in Australia*. Bureau of Immigration Research (BIR).

Rochefort, D. A., & Cobb, R. W. (1993). Problem definition, agenda access, and policy choice. *Policy Studies Journal, 21*(1), 56–71.

Simon, H. A. (1957). *Administrative behaviour: A Study of decision making processes in administrative organisation* (2nd ed.). Collier/Macmillan.

van Beinum, H. (1998). On the practice of action research. *Concepts and Transformations, 3*(1/2), 1–29.

Weiss, J. A. (1989). The powers of problem definition: The case of government paperwork. *Policy Sciences, 22*, 97–121.

Yanow, D. (1987). Towards a policy culture approach to implementation. *Policy Studies Review, 7*(1), 103–115.

CHAPTER 3

From the Local to the International: Domestic Politics and International Advocacy 1990–1992

Narrated by Eileen Pittaway

Becoming an 'Expert'

It is amazing how quickly one can become an expert! In July 1990, a group meeting of experts on refugee and displaced women and children was convened at the United Nations, Vienna, by the Division for the Advancement of Women (DAW), a secretariat of the Commission on the Status of Women (CSW). The purpose of the Expert Group Meeting was to provide substantive input into a paper to be submitted to the thirty-fifth session of CSW in March 1991, which would have 'refugee and displaced women and children' as its priority theme. The outcome of this session would be presented for ratification by member states to the United Nations General Assembly (UNGA), in September 1991. Participants comprised 60 representatives from the United Nations High Commissioner for Refugees (UNHCR), donor and host governments, including Australia, academics and non-government organisations (NGOs). There were no refugee women invited to take part in the meeting or the drafting of the document.

Even though I was still writing up our research findings and the report had not yet been published, a request was conveyed through the Australian Office of UNHCR that I be invited to the meeting, and an invitation was extended by the Director of DAW to present a short paper on the findings of the research to date. The trip was partly financed by

E. Pittaway and L. A. Bartolomei,
Only Rape! Human Rights and Gender Equality for Refugee Women,
Sustainable Development Goals Series,
https://doi.org/10.1007/978-981-19-0916-0_3

ANCCORW and the Australian Council for Overseas Aid (ACOA), and partly self-funded. We hoped that our input into this meeting would not only contribute to outcomes at the international level but would also influence the direction of the Australian Government's policy on refugee women.

Thrown in the Deep End

On arrival in Vienna, I was feeling nervous and totally unprepared. It was my first UN meeting, and I was not sure what to expect, where to sit or what my role would be. I was very aware that the paper I was presenting challenged the conventional wisdom about refugee women and torture. Imagine my surprise when, immediately upon my arrival at registration, I was informed that I had been appointed as the rapporteur, with responsibility for summarising the meeting outcomes (I had no idea what the role of a rapporteur even was). I was given a seat on the stage and a slot in the closing ceremony to make my report. When I asked what the expectation was from the report, I was unhelpfully told, 'Oh you are an academic, you know what to do! We know you write well, and Australia is a neutral county'. At the end of the meeting, I asked if my report was satisfactory. 'Oh yes', I was told, 'we like it. Refreshingly honest, and different to the 'diplomatic speak' we often get'.

My presentation at the meeting described the ANCCORW project, the preliminary findings on resettlement services and also the high occurrence of rape and sexual abuse experienced by refugee women. I was so nervous that as I stood at the lectern, I could not stop my left leg from shaking! However, as I looked around the audience, I saw people leaning forward, engaged, and many heads nodding in agreement. The findings were accepted by the meeting, and many experts commented that they confirmed trends which were emerging in other resettlement countries and supported evidence of rape and sexual abuse in refugee sites. At the end of the meeting, a report, with recommendations, including acknowledging and addressing rape and sexual abuse, was presented to the chair for submission to CSW 35.

My attendance at this meeting had serious ramifications for my relationship with government officials in Australia. My participation had been arranged by UNHCR Australia, and the Government did not know that I was attending. There had been no official briefing and I was not aware who would be representing the Australian Government. I was not even aware of the protocols for relationships between government representatives and the non-government sector at such meetings. The two women from the Australian Government, who were present as observers, were surprised to find me at the meeting, especially when I was appointed rapporteur.

While at the meeting, I made contacts which enabled me to become part of a network of the major players in the international refugee domain. The UNHCR Senior Co-ordinator for Refugee Women, Anne Brazeau, was very impressed with the work of ANCCORW, and an effective working relationship was developed. She suggested that I should join the international CSW (NGO) caucus working to take refugee issues to the Fourth World Conference on Women, planned for 1995. Through a contact at the United States Mission (embassy), she invited me to dinner with the US Ambassador to the UN to discuss strategy with several influential guests.

Over the course of the CSW meeting, I was also invited to a breakfast briefing with the Australian Ambassador, lunch with the Ambassador's wife and drinks in the embassy mess. The unwritten agenda of these meetings was to find out what was going to be reported from the NGO sector, and to attempt to influence the position I would be taking. I was reminded how generous our refugee policy was, how good our service provision was for refugees in Australia, and of the positive response of the Government to refugee women. They were obviously not pleased with some of the things I had said in my presentation. However, I received a warmer reception from some individuals than others; one of the staff at the mission provided a number of contacts in Canberra that she thought would be useful to ANCCORW, and subsequently facilitated my attendance at further international meetings.

Other positive outcomes from that meeting in Vienna were that the validity of the work being undertaken by ANCCORW was acknowledged. There was considerable interest in the statistics about rape and sexual torture, which confirmed anecdotal stories from around the world. Ann Brazeau was invited by the Australian Government to launch my report at a BIR conference in Australia. The main lessons I learned were about the complexity of working in 'the UN system', and especially the need to study how to work within the protocols of diplomatic nicety, and to understand the politics of policy-making.

No More 'Mr Nice Guys'

On my return, I was contacted by the Australian Government to discuss the progress of the ANCCORW project with the head of resettlement services. They had received reports from Vienna and expressed concern that the research report might criticise service provision. They suggested that if they were not pleased with the report, the invitation to Ann Brazeau would be cancelled.

The research report was completed, and the first draft sent for consideration to BIR in December 1990. It identified 'major procedural and administrative problems of service provision, both government and non-government'. The women interviewees had identified seven priority areas as essential for successful resettlement in Australia. Recommendations were made for the provision of new and amended services to address these special needs.

The report was forwarded to DILGEA Resettlement Branch, Canberra, for comment. The reaction was extremely negative. They took exception to the criticism of services and demanded BIR withhold the report. I was called to Canberra for discussion with the heads of the departments criticised in the report. Pressure was put on me to significantly alter some sections and withhold some of the findings. The invitation to Ann Brazeau to launch the report was cancelled. However, BIR was annoyed at what they perceived as interference by DILGEA in their sponsored research. In February 1991, the final draft of the report was completed, without major amendments, and submitted to BIR to be refereed. I was offered the newly funded GIA position with ANCCORW to follow up the research findings.

In March 1991, CSW held its annual meeting, attended by Australian Government delegates. A key agenda item was to consider and adopt the paper on which we had worked at the Expert Group Meeting, for the United Nations General Assembly (UNGA). Although it is traditional that when NGOs have had input into the proceedings of UN meetings, they are invited to join their government delegation to attend, the Government declined to invite me and refused to include me in any of the discussions. While they believed they had prevented me attending, I was directly invited to attend by the Director of the CSW, because of my input into the paper under discussion. Seeing me at the meeting, the government representatives actively, and quite rudely, excluded me from their delegation on the first day. Other participants at the meeting expressed surprise at their behaviour, because of the prominent role I had taken at the expert group meeting. Excluded from participating as part of the Australian delegation, I planned to make the best of the meeting from the side-lines… but fate intervened.

The Power of 'Being There'

During the first day, a Norwegian friend[1] and I were having coffee when the US Ambassador to the UN, whom we had met at the previous meeting, greeted us and asked how we were. We expressed our disappointment at not being able to get tickets for the opera while we were in Vienna. 'Oh', the Ambassador replied, 'I have a box tonight with only my aide. Why don't you join me?'. The concert hall at the Vienna Opera house is a grand semi-circular edifice with many private boxes over several floors. It is the sort of place one goes to be seen, as well as to enjoy the theatre. It was with great delight that we joined the Ambassador in her box and, before the opera began, had great fun hanging over the balcony, watching people, whilst sipping champagne. Imagine our satisfaction when, in the back row of the stalls below us, we saw the Australian delegation and happily waved down to them. We then sat back and enjoyed the opera. When we returned to the hotel, a big surprise awaited me. A very polite note apologising that the Australian delegation had not contacted me and inviting me to breakfast with the Australian Ambassador the next day. A bundle of briefing notes was included, and I was invited to join them as part of the delegation. I declined. I did, however, attend the lunch for ANCCORW hosted by the ambassador's wife. Relationships were very cordial, and best of all was the news that the Government had reconsidered its decision to accept the report and that Anne Brazeau was once again invited to launch it. This was all based on the obviously intimate relationship I had with the US Ambassador to the UN! I had been seen at a US-hosted dinner the previous meeting and now I was with her at the opera. What was I saying and what influence could I have? This casual and serendipitous event changed my status from being marginalised by the Government to being one of the privileged insiders.

There were further benefits from just 'being there' in Vienna. In her closing speech, the Australian Government representative stated, among other things, that when women at risk (WaR) arrive in Australia, they and their families receive additional settlement services to facilitate a successful adjustment to their new environment, and that refugee women were well serviced by the torture and trauma counselling services which had been set up in Australia to assist refugees reconstructing their lives. This was patently untrue. At that time, there was only one small torture and trauma service based in Sydney, and the majority of their clients were men. Refugee participants in the research project had complained that there was nowhere for them to seek help. Furthermore, the staff of the service were somewhat dismissive of the reports of sexual abuse experienced by the women.

The government delegate also announced the upcoming conference at which my report would be launched. With a smile, I asked for a copy of her speech and immediately faxed it (no email then!) to my extensive networks in Australia, with a request that they take action to pressure the government to fulfil its word. Government offices were flooded with faxes from NGOs around Australia requesting meetings with department heads and politicians. Within three months, torture and trauma services were established in all states receiving refugees, offering services to both men and women. I am not for one moment pretending that this happened just because of my fax. However, without access to that speech, which claimed the services already existed, the fight would have been much harder. My work at the international level had not been planned nor anticipated, but it had helped to give ANCCORW a very new and very small NGO a high profile at the national level in a short space of time.

On the way back to Australia from the CSW meeting in Vienna, I visited refugee camps in Hong Kong as the guest of International Social Services, an NGO working in refugee camps. The purpose of the visit was to investigate failure of local UNHCR officials and Australian officials based at the Australian post in Hong Kong to implement the WaR Program. The women I interviewed reported horrendous stories similar to those shared by the women in Australia.

Kip's Story—Sex for Survival

We have to queue for food in the hot sun with no shade for hours. Last Monday, when I got to the bench to collect food for my family, husband and two small children, a soldier said, 'Come round the back of the hut and have sex with me and I will give you double!' I said, 'No I will not do that'. He then pushed me away, saying 'Then today you go hungry'. Later he came to our shelter, pushed my husband outside, locked the door and violently raped me, in front of the children. When he had finished, he laughed and said, if you had done that when I asked you, you would have had double food – now you have nothing'. My husband is so ashamed he has not spoken since, and the children are terrified. I still have to go every day to get the food but the soldier ignores me (Pittaway, 1991).

In a four-day visit, I visited five refugee camps and detention centres and consulted with several NGO and middle-level UNHCR officials. It was established that none of the grassroots workers who were in a position to identify 'women at risk' knew about the Program. However, many identified women whom they considered eligible for the Program. I gave them information to assist with referring these women to 'determination officers', those officials who made the decisions about which asylum seekers were granted refugee status, and which refugee women were given access to the WaR Program. However, despite the guidelines, a structure was not in place for this to happen. There was no treatment or support for women who had been raped or sexually abused prior to arrival, and little protection within the camps. Women and service providers in Hong Kong indicated that, often, camp security staff perpetrated abuses within the camps. Officials working at senior levels, such as Australian post officials and UNHCR middle-level management, were either not aware of the Program or were not passing information down the line. Despite instructions from Canberra, it appeared that these senior staff were indifferent to the Program.

On my return, despite having announced the launch of the report, the Australian Government made one last effort to get me to change it.

Last Ditch Government Attempts to Influence the Report

In April 1991, unofficial attempts were made to have ANCCORW reconsider its report. A meeting was held in Sydney between a senior bureaucrat and myself, with two students on placement. He arrived inebriated following a Prime Ministerial function and the ensuing conversation swung between 'let's be pals' and scarcely-veiled threats about what he might do if the report findings went forward unchanged. One of the comments was that 'we both know that the GIA Scheme is just a sop to ethnic communities'. He threatened that if we continued with our 'ridiculous assertion' about the extent of torture and trauma, he would take action to discredit the whole research. When I refused to agree to his suggestions, he became aggressive and said, 'Listen, you just stop this nonsense. The stories aren't true, our agents (Australian Security Intelligence Organisation) just made-up stories about refugee women being raped by pirates and fed it to the press in order to get national acceptance of the Vietnamese program'. (The rape of Vietnamese women is well documented through interviews with women held in the camps prior to resettling in Australia, Canada, and other countries, UNHCR, 1990). Also, in full view of the students, he fondled my knee and thigh during the meeting. It would appear that on return to Canberra, he perhaps recognised that the meeting had been a tactical error and that some of his statements may have been indiscreet, to say the least. The following week, one of his senior staff came to Sydney to 'smooth things over' and set things right.

I was at this time documenting the process as part of my doctoral thesis. When I reported what had happened to staff at the university, no support was offered to me. Instead, I was told that I could not include it in my PhD in case the university was sued. When I insisted, the university legal department instructed my markers to sign an agreement not to disclose the content of that particular chapter, and it was removed from the released version of my PhD for a period of 20 years.

The unchanged report on the resettlement experience of refugee women, *Refugee Women—Still at Risk in Australia,* was launched in May 1991 at a very successful international conference, which attracted 500 participants and wide media coverage. Over the next decade, it informed major changes in refugee policy relating to refugee women and girls.

In July 1991, now employed as a community worker with ANCCORW, I supervised two mature-aged students, Sylvia Winton and Diane Hayes, to undertake follow-up research for ANCCORW, to identify the barriers experienced by women in relation to the WaR Program. While Diane followed up with women in Brisbane and Sydney, Sylvia travelled to Thailand and Hong Kong, at her own expense, to research the issues in refugee camps. Meanwhile, Velvy Holden and I travelled to camps in the Philippines. In all sites, problems were identified in the transmission of information about the program, from embassies and consulates to field staff working with refugees. Field workers were consistently surprised to learn about the WaR Program and indicated that many of the women with whom they worked would probably be eligible. Lack of gender sensitivity was also a major problem, with both UNHCR staff and officials from Australian posts. Women reported that they did not disclose information about sexual abuse, which would have established their eligibility for the Program, to the male determining officers. Most importantly, major problems existed in the interpretation of the eligibility criteria for the Program. For example, one settlement officer in Hong Kong reported that he did not believe that there were any women at risk in a refugee camp in Hong Kong, as women were free to marry under Hong Kong law. The implication was that if women were married they were safe, and that if they were without 'male protection' it was through their own choice. Problems such as these resulted in eligible women not being recognised as potential applicants or being submitted to overseas posts for approval as WaR (Hayes & Winton, 1991, pp.13–18). UNHCR officials reported that in some places, it was quicker and easier to get a woman in a situation of danger removed to other resettlement countries through the normal channels than to wait for Australia to use this supposed 'fast tracking' system. Little concern had been shown by the Government about the apparent inability of posts to fill the annual quota.

Based on these research findings, ANCCORW submitted a report on the WaR Program to the Australian Government, including a range of recommendations to address these problems, such as NGO workers being given the power to refer women to the WaR Program, and more women being recruited and trained as determining officers at Australian posts. Some note was taken of the recommendations, and in 1992, ANCCORW was involved with DILGEA in designing new information brochures for staff and in providing gender training to officials going to work in overseas posts. Training for existing staff was provided on an ad hoc basis. Still hopeful, still naïve, I thought that the problem had been addressed (Pittaway, 2001).

Advocacy and Lobbying Activities for the WaR Program

The report on the WaR Program was used extensively as a lobbying document by ANCCORW and its member agencies. At a consultative meeting about refugee intake in 1992, the then Minister pledged to double the number of cases in the quota for the WaR Program and commented, 'I have such a strong personal commitment to this program that people will be bloody well be crucified in Canberra if it doesn't work!!' In fact, the quota was lifted to 200 places per year, with the promise that, if more women were identified through UNHCR or Australian posts, there would be no restriction on their entry. Despite initial jubilation at these outcomes, nothing changed. There were major problems in the Department (DILGEA), ending in a closed court case and the dismissal of two senior staff who had been involved in managing the Program, and this slowed the process right down. At the same time, the Minister for Immigration resigned to take up advocacy work on behalf of refugees.

The new Minister was much more hard-line. He expressed public opposition to RCOA and stated that he would not fund them, and he would like to see them closed because of their advocacy work. The Minister also took issue with the head of BIMPR (formally BIR), who had been supportive of ANCCORW. The basis for this conflict was the continued independence and autonomy that BIMPR demanded as a research body. It is reported that the Minister stated at a departmental meeting that the role of those he either directed or funded was to follow and not to question instruction. Subsequently, the head of BIMPR was removed from his position, and the Bureau was downsized in 1994. This is one of many examples we encountered of the power of an individual to control policy direction. Despite the Minister's resistance to much of the direction being suggested by the community sector at this time, he was not opposed to the WaR Program and stated that he wanted it to succeed. It was a politically palatable cause, acceptable to the public and useful as a counter argument to criticism of other aspects of Australian refugee policy.

Smoke and Mirrors—The Failure of a Process

The number of visas awarded for the WaR category was carefully monitored by ANCCORW over the following months. By early 1993, it became obvious that, despite the steps taken by the Government to address the implementation problems, train overseas posts and disseminate more comprehensive information, women were still not being selected for the program. The numbers of WaR entering Australia slowly began to increase, but only 64 of the new quota of 200 positions were filled in 1992/1993 and few additional settlement services were provided to meet their needs. Despite the evidence-based research, the advocacy, the promises made and amended policy, nothing changed. We were furious!

Conquering the United Nations

When it became obvious that change was not going to come about at a national level, the ANCCORW board decided we would have to progress to the international arena. This took me on an incredibly steep learning curve. In order to advocate effectively, I needed a firm grasp of the UN system, as this is where major decisions over the lives of refugees are made. This included important international frameworks and agreements, refugee and human rights law, conventions and policies. One thing I had already absorbed was the serendipitous nature of how the system works, and how some of the most important things achieved came about from processes far outside the expected norms of policy analysis and evaluation. Who would have guessed the impact that being seen at the Vienna Opera with a US Ambassador would have on national refugee policy?

The UN is arguably the most complex and cumbersome bureaucracy in the world. An important part of the work undertaken was learning the complex structure and language peculiar to the different 'organs' (sections) of the organisation. Understanding the process is daunting, and unless you have been to the UN yourself, you will need to understand something about the structure and workings of the system in order to follow this convoluted and complex story.

The first thing to understand is that the UN is akin to the bureaucratic framework of a world government, but without many of the powers. The 'parliamentarians' in this 'world government' are the representatives of member states (governments). The UN produces international law, but this is not enforceable at an international level. It depends on moral imperative and the political will of member states to incorporate international laws into their domestic legal frameworks. The UN has no power to force member states to implement law and policy (it has a peace keeping force but no military arm), but it can occasionally impose sanctions. While this lack of executive power is often seen as a major failing, it can also be argued that it is a major strength. A rogue world government would be disastrous! Much criticism is levelled at the various UN bodies for being ineffective, and the structure of the Security Council, with five nations having power of veto over all decision-making, is particularly problematic. It is good to remind ourselves that the various UN bodies, flawed as some may be, also do an immense amount of good, and we argue that the world would be in an even worse state than it is if it did not exist.

The UN has six working languages for which they provide simultaneous translation at all official meetings: they are Arabic, Chinese, English, French, Russian and Spanish. However, English is the lingua franca of the majority of meetings, and as English is my first language, I was confident that I would be able to comfortably follow proceedings. How wrong I was! To participate fully in meetings at the UN, one has to become familiar with a coded version of English which we have named 'UN-ese'. To be accepted as an 'insider', it is necessary to speak this 'dialect'; not to know it places you firmly on the outside. When is a conclusion not a conclusion? What is the difference between a conclusion and a resolution? What is the distinction between signing and ratifying? What is a 'non-paper' and what differentiates a formal from an informal meeting? How does one 'raise a flag' in a formal meeting room—and what does that mean? Who has the right to speak and when? What is a 'language' session? How do the complicated microphone systems which carry the translation work? It is important to know that in Geneva, UNHCR is referred to as HCR, and that in UN meetings one never 'speaks', one 'makes an intervention'! Just to complicate matters, the language changes between UN bodies, not to mention the forms of greetings. In New York, a hand-shake will do, although kissing has become more acceptable. In Geneva, even relative strangers get a two-kiss greeting, but you know you have arrived when you get three kisses. COVID-19 has really confused the ability to know one's place in the UN social hierarchy! Language expression is also very important: seldom, if ever, are direct or strong criticisms made, or member states spoken about negatively. One has to learn the art of diplomatic niceties, and how to deliver sharp barbs veiled in silk. This may sound trivial, but it is not. Part of our work over the years has been to provide as much information and training as possible to colleagues going to UN meetings for the first time, to make sure that they do not feel totally alienated, and that they are effective in their advocacy.

The Structure of the United Nations

The charter of the UN established six 'principal organs'. They are the General Assembly (UNGA), the Security Council (UNSC), the Economic and Social Council (ECOSOC), the Trusteeship Council, the International Court of Justice (ICJ) and the Secretariat. In addition, there are a number of specialised agencies, which include UNHCR. UNHCR is the only agency which has the responsibility to directly receive funding and to provide services to refugees. Other agencies are purely law and policy focused. The critical role of UNHCR has to be recognised when analysing the formulation of policy, both internationally and nationally (Fig. 3.1).

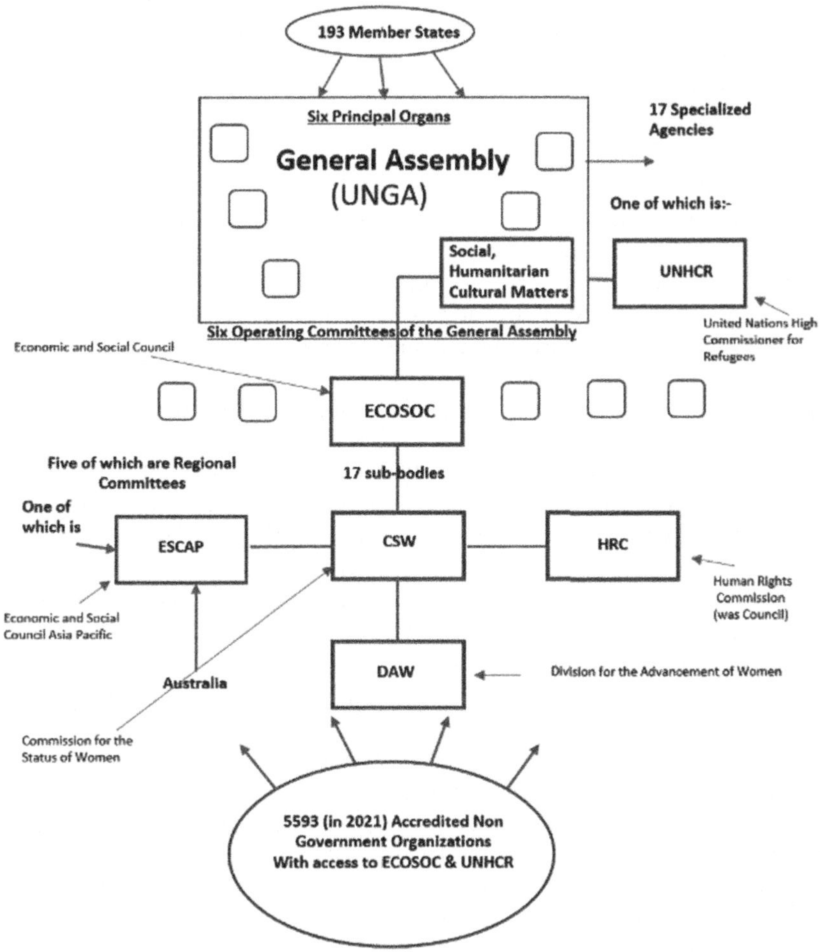

Fig. 3.1 Structure of the United Nations

The UN General Assembly (UNGA) is the main organ of the UN and is linked to all other organs. It has wide-ranging powers, and recommendations made in other organs and agencies are taken to the regular sessions of the UNGA for discussion and adoption. To deal with the quantity of work at each UNGA meeting, six separate committees sit to debate specific issues. The work covered in this story is considered in the third committee, ECOSOC, which covers social, humanitarian and cultural matters.

ECOSOC was established to coordinate the economic and social work of the UN and its specialised agencies, known as the UN Family of Organizations (UN, 1998). The council has six functional commissions, including the Commission on the Status of Women (CSW) which, among other areas of responsibility, organises the World Conferences on Women that play a critical role in this story. Voting rights in CSW are rotational, with only forty countries holding this right at any given time. ECOSOC has divided the world into five regions: Europe, including North America and Canada; Latin America and the Caribbean; Africa; West Asia; and the Asia–Pacific, and has appointed economic and social councils in each of these. Australia is geographically situated in the Asia–Pacific Region, which extends from Pakistan to New Zealand and includes Japan. The local secretariat is the Economic and Social Council of Asia–Pacific (ESCAP), located in Bangkok. Although geographically situated in this region, Australia has been placed historically in the European region because of socioeconomic and cultural ties. Increasingly, and for the Fourth World Conference on Women held in 1995, it was more aligned to the Asia–Pacific region.

From time to time, ECOSOC convenes large and significant 'World Conferences' to focus attention on a current global issue. These involve a number of preparatory meetings held around the world and finish at a major global event hosted by one member state. In 1975, CSW supported by UNGA convened a World Conference on Women in Mexico to address the emerging concern about gender inequality across the world. This was followed by a conference in Copenhagen in 1980, in Nairobi in 1985 and in Beijing in 1995. The documents and reports from these meetings were amended at successive meetings of CSW for four years prior to each conference. The aim of the Fourth World Conference on Women was the finalisation and endorsement of the 'Beijing Platform for Action' (BPFA), a blueprint for the improvement of the status of women to the year 2000 and beyond. There have been many other World Conferences, but in the twenty-first century, these have been scaled down due to the threat of terrorist attacks.

The term 'member states' is used to describe governments which are members of the UN. There are currently 193 member states. A number of NGOs are accredited to attend ECOSOC meetings, with an increasing trend for this level of civil society involvement in activities of the UN system. The number of accredited NGOs has grown from 41 granted consultative status in 1948, to 1,350 in 1998, when ANCCORW received consultative status, to 5,593 in 2021. Invitations can be extended to non-accredited NGOs by the meeting convenors. The role played by NGO lobbyists is acknowledged by the UN agencies and by other institutions.

The UN agencies, such as CSW and UNHCR, hold annual, high-level, government meetings at which they prepare recommendations to UNGA. UN agencies often convene 'expert group meetings' of international specialists to prepare input into the annual meetings, such as the one I attended in preparation for CSW 35. Meetings of UN agencies have differing status, determined by the level of government representation. Most agency meetings are defined as 'high-level government' meetings, and it is expected that a senior bureaucrat attends these. The next level up are ministerial meetings, and the highest level are heads of government meetings.

How UN Law and Policy is Written

The process of producing law and policy in these agencies is long, slow and complicated. Draft documents prepared by UN staff are sent to governments for consideration. A series of intergovernmental meetings are then held, sometimes lasting for years. During that time, the draft documents are continually amended and refined until the final meeting on the issue, when governments are asked to sign the documents. This means that some governments sign off, i.e. approve, documents at the UN, but never ratify them by incorporating them into their legal systems at a domestic level. It is therefore often argued that UN conventions and protocols are of little real value, because there are no mechanisms in place to enforce them. It is up to individual member states to enshrine UN conventions in their domestic law, and to uphold them. For countries with a Bill of Rights, UN conventions automatically become part of their domestic law once the member state has ratified them, but this still does not guarantee that they are upheld. In countries like Australia, however, which is one of the few developed nations which does not have a Bill of Rights, this is not an automatic process, and governments choose which principles they incorporate into domestic legislation. Governments can therefore sign, ratify and then ignore these documents with relative impunity. There is, however, a moral imperative in human rights law, and few developed countries like to be publicly accused of transgressions.

In CSW and the UN Human Rights Council (UNHRC), amendment is through a process referred to as 'square brackets' and 'language' sessions. At each meeting, representatives from member states discuss the document paragraph by paragraph and agree on certain language. If agreement cannot be reached on parts of the document, they are placed in 'square brackets' until the following meeting. The task of the final meeting is the resolution of the language still in square brackets. NGOs and civil society have no formal role in this process but are usually allowed a very short intervention for each agenda item. However, they can prepare suggested language, and advocate and lobby for governments to accept it. This happens at informal meetings with member states, in corridors outside meetings, at cocktail parties and coffee bars. These bodies also produce resolutions, or 'soft laws' at major meetings, which guide states on the implementation of international laws and conventions.

UNHCR does not use the square brackets process. Instead, they have a series of 'Standing Committee' meetings each year, and decisions are made by member states in closed sessions rather than through debate over square brackets. 'Resolutions' are referred to as 'Conclusions'. Thus, NGOs have an even more difficult time in getting their language included. They are allowed to observe these meetings and to make short joint interventions which are coordinated by the International Council of Voluntary Agencies (ICVA), a Geneva-based group which represents the majority of INGOs working with refugees and Refugee Councils globally. ICVA and UNHCR hold 'UNHCR- NGO Consultations' on an annual basis, and the outcomes of these are fed into the Standing Committee meetings. With NGOs lacking many formal channels, an enormous amount of 'back door' advocacy takes place at UNHCR, and the good news is that it is often successful. The 'language' written by advocacy groups can be found throughout UN legal and policy documents. However, to fully and effectively participate, NGOs need to be trained in this complex process and be familiar with UN language.

Why is the United Nations so Important to This Story?

It was not until the Nairobi Women's Conference, convened by the United Nations Commission on the Status of Women (CSW) in 1985, that sexual violence and the vulnerability of refugee women and girls were noted as an area of serious concern, and member states and UN bodies, in particular CSW and UNHCR, made commitments to address them. Over the next two decades, it became one of the key issues addressed in the 1995 UN Beijing Women's Conference, then in the formulation of the Statute for the International Criminal Court (ICC) and at the UN Security Council. ANCCORW joined with colleagues from within Australia and across the world in an international collaboration to bring the issue of the lack of effective protection experienced by many refugee women and girls to these UN meetings.

The mandate of UNHCR is to provide international legal protection for ALL refugees, returnees, stateless persons and, in certain circumstances, Internally Displaced Persons (IDPs), and to find durable solutions for them. It works to provide both prevention and response mechanisms to address the issue of the rape and sexual abuse of refugee women and girls. Despite the recognition within the UN system of this issue for many years, refugee women and girls are consistently let down by the international protection system. One main reason for this is that the 1951 Refugee Convention has only been signed and ratified by 145 of the 193 member states, and most of those who have not signed are the countries hosting the majority of the world's refugees. The refugees in these countries, which are primarily developing nations, many with dubious human rights records, are therefore not protected by international refugee and human rights law. Many developing nations host refugee populations, while with few exceptions, developed nations including Australia do everything they can to prevent refugees claiming asylum on their shores.

A Recipe for Successful Advocacy

As well as building a theoretical framework, our work required us to build a model for successful international advocacy. The actions in each circle are equally important and most have to be undertaken simultaneously. Too often, advocacy is missing an important component and fails. International advocacy is time-consuming and expensive. Funding is hard to find. Importantly, it is a long-term process. Unless there is a catastrophic trigger for change, it can take years. The important point is that it can work. Most major social changes come about because of the work of advocates, not because politicians are intrinsically good and caring people.

In this early cycle of research, we had just begun learning how to advocate effectively. One key lesson was the importance of building the right networks: the power of who you know, and—as we saw in Vienna—who people think you know! Advocacy is a collective endeavour and can seldom be done alone. Another lesson, gained from our use of the data collected in Hong Kong and Manila, was the need for good evidence-based research. Other understandings were beginning to emerge. We quickly found that we had to become familiar with the complex UN systems and the drafting of effective language for advocacy. The importance of getting the voices of refugees to key meetings was becoming more obvious, although still at this stage, we mainly did this through case studies and multiple quotes in all of our reports. Finally, the need for a stronger theoretical framework was self-evident (Fig 3.2).

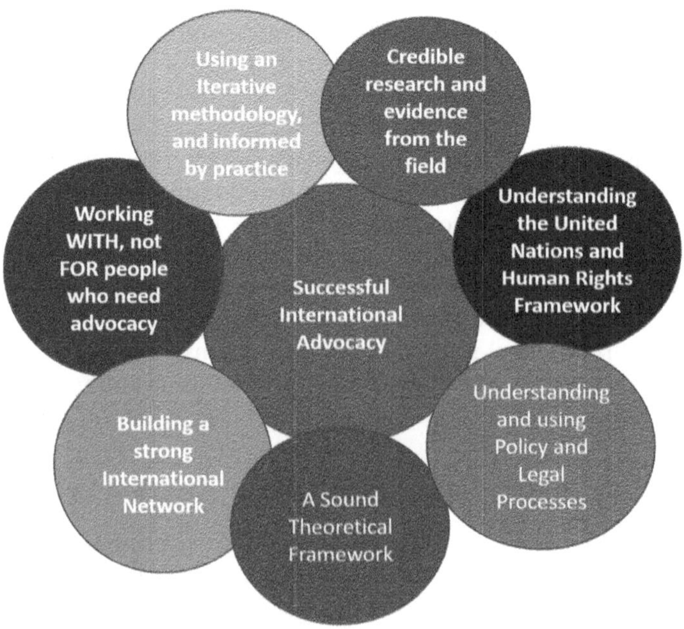

Fig. 3.2 A recipe for successful advocacy

What Theoretical Analysis Taught Us

Applying the political paradigm to analyse the policy process in which we had been caught up was enormously helpful to my understanding of the repeated policy failures and the role of various actors. At times, we struggled to understand the reaction to our research findings. It was very apparent that, at both national and international levels, the process was inherently politically driven, but the 'politics' were convoluted and often lacked any apparent logic. For example, the very adverse reaction of the Australian Government to the ANCCORW research suggested that, although they had funded it, they had not expected, nor were prepared to accept, adverse findings.

Scholars using the political paradigm contended that neat and rational links between policy and politics do not exist, problems are unpredictable, and issues are not addressed until they impact on the policy process (Barrett & Fudge, 1981). This was patently obvious in the reaction of government officials to my participation in the two meetings in Vienna.

Considine (1994) states that those responsible for the implementation of policy work frequently transform or subvert the policies handed down to them. The way in which they interpret the world predetermines the way in which policy is developed and interpreted. This was evident in the refugee camps in Hong Kong and Manila, where the rhetoric of the Minister had little impact at ground level. The role of the 'street level bureaucrat', described by Lipsky (1980), explains the power of people working at the coal-face, far from the elites who formulate policy, to interpret and implement policy as they see fit. Their actions are seldom monitored or evaluated, and their distance from the decision-making body allows wide interpretation, or even ignoring, of policy directives. The response by officials overseas to the implementation of the WaR Program was a prime example. It made us question if the Minister really had been committed to the WaR Program, or if it had been a tokenistic response to appease critics and to enhance Australia's reputation on the international stage.

Above all, what we learned is that the policy process is indeed profane and depends on the subjective interpretation of different actors. The 'on again-off again' invitation to launch the report was a classic example. ANCCORW found themselves caught up in a number of battles—with government officials, with RCOA and with service providers who did not agree with our position—and although these were about policy, they were very much political contests. The concepts of the sacred and the profane were invaluable in assisting me to navigate these minefields. If ANCCORW colleagues and I, as an academic, had limited ourselves to the sacred view of the policy process, we would have experienced barriers at every level, preventing us from exerting upward influence. However, because of our ever expanding networks and supports, we placed ourselves firmly in the centre of the profane circle of influence and used our informal power and knowledge to challenge the system.

Conclusion

If you notice a few chronological discrepancies in the story and discussion, remember our warning in the Prologue about the complexity of this story. At times, action in Australia and action overseas were so intertwined, that to tell them separately would be to lose the pattern of the story but telling them in the exact order in which they happened would lose an important point.

Although we had come a long way, our analysis still did not address the central question of why people apparently did not believe, or care about, the abuse of refugee women? It gave us knowledge of *how* they subverted the system, but not *why*. This is what we needed to discover before we could proceed. Clues came from the fact that the DILGEA bureaucrat with whom I had met had assumed he could get away with blatant sexual harassment. How were we going to fight to end the sexual abuse of refugee women when it was accepted as the norm in the rest of the world? How telling was the reaction of the university where I did my PhD (not UNSW), whose response to that story was to protect themselves from possible legal action rather than supporting me and confronting the perpetrator? 'Politics' alone could not explain these attitudes. We needed to better understand what influences the way decision-makers define problems and solutions.

This realisation guided our planning of the next research cycle. The political paradigm provided only some answers, and we were beginning to identify the gaps. One thing that was becoming obvious was the lack of application of a human rights framework, except as a political trope. A feminist analysis of the weight of the patriarchy was also missing, but I did not yet understand the full implications of this. It was not until we moved into the next stage of the story, and I delved deeper into the theoretical analysis, that things finally fell into place. Theories in the critical structuralist paradigm provided answers to these questions and continue to do so. But rather than providing neat solutions, they posed problems bigger than we had ever imagined!

Note

1. Eleanor Brenna, Danish Refugee Council.

References

Barrett, S., & Fudge, C. (1981). Examining the policy-action relationship. In S. Barrett & C. Fudge (Eds.), *Policy and action*. Methuen.

Considine, M. (1994). *Public policy: A critical approach*. Macmillan.

Hayes, D., & Winton, S. (1991). *An evaluation of the women at risk scheme*. ANCCORW.

Lipsky, M. (1980). *Street-level bureaucracy*. Russell Sage.

Pittaway, E. (1991). *Field notes*.

Pittaway, E. (2001). *Only rape!: An examination of the power of ideology and discourse in the policy process with a focus on policy pertaining to refugee women*. PhD thesis, UTS, Sydney.

UN. (1998). *The structure of the United Nations*. Retrieved October 22, 2021, from http://www.geocities.com/CapitolHill/Lobby/344//structure.num

UNHCR. (1990). *Statement by UN high commissioner for refugees to the extraordinary session of the executive committee*. UNHCR.

CHAPTER 4

Rape as a War Crime: The Road to the 4th World Conference, Beijing 1993–1995

Narrated by Eileen Pittaway

> *'I cannot grant this woman refugee status on the grounds of rape alone – as rape is the common experience of women everywhere!'* (USA Judge in Castel 1992).

At this stage of the story, at the beginning of 1993, the ANCCORW team felt disappointed, overwhelmed and hopeless. We had researched, analysed and played the politics well. We had hoped to bring about major change and had failed. Faced with entrenched ideological resistance, ignorance, tokenistic policy-making and lack of an implementation strategy for the changes we had fought so hard to make, we were unsure where to go next. The research findings identified the problems inherent in the Women at Risk (WaR) Program, but our analysis to date had not explained the underlying causes. Fortunately, there was a light at the end of the tunnel.

I revisited the research data searching for something I may have missed, and began to dig more deeply into feminist literature, critical discourse analysis and theories of power as potential sources for unpacking the issues. The root of the problem appeared to be *why* the rape and abuse experienced by so many refugee women was not recognised and acknowledged as 'torture and trauma' and therefore as grounds for protection under the WaR Program. The re-evaluation of the data

E. Pittaway and L. A. Bartolomei,
Only Rape! Human Rights and Gender Equality for Refugee Women,
Sustainable Development Goals Series,
https://doi.org/10.1007/978-981-19-0916-0_4

through a different theoretical lens strongly suggested that the issue of interpretation of the women's experiences, shaped by ideological positions, was the crux of the matter. Until this was resolved, other changes would be purely cosmetic.

'Only Rape'—The Lightbulb Moment

It was noted that seven of twenty-two senior male officials interviewed for the project (in Thailand, Hong Kong and the Philippines) had used the same phrase when asked if they considered rape and sexual violence as grounds for referring women to the WaR Program. The phrase was *'only rape'*, implying that rape, or the likelihood of being raped, was not sufficient grounds for considering a woman for this visa. Their argument was that if the only thing a woman complained about was rape and sexual abuse, then she could not possibly be considered as a woman at risk. An official at Whitehead camp, Hong Kong, commented, 'if rape is the only criteria, I could send you most of the women in this camp. It happens all the time, especially to the young single women, and we can't do much about it'. A UNHCR official in Hong Kong stated that rape was not grounds for refugee status, so therefore it could not be grounds for the WaR category, and that to qualify for this program a woman had to be experiencing extreme forms of violence, not only rape. An official in *Site One Refugee Camp*, Thailand, said that rape was so common it could not be considered as grounds for recognition, and, anyway, it was how women got some extra food (from the guards who raped them), and was therefore hardly likely to be classified as 'extreme danger'. The most damning comment, from *Site Two Refugee Camp*, Thailand, was that 'often what happened was not really rape anyway, because some women 'exploited' their sexuality within the camp system to get favours from the guards'. A male camp manager at *Site One* stated that, because many women had already been raped by the time they reached the camps, rape in the camps wasn't a problem, and the final comment, from the manager of a camp in Manila, was that 'it happens so often to these women that they get used to it, sort of expect it, and they don't see it as violence like being beaten up or tortured' (Pittaway, 2001).

Despite much rhetoric about protecting refugee women, there was an unwillingness or inability by many men in positions of influence to accept rape and sexual torture in conflict and refugee situations as a major problem. This has been well documented internationally (Bhabha, 1993; Castel, 1992; Hayes & Winton, 1991; Mahud, 1996). An incident at a meeting in Sydney reinforced this insight. When informed of a case involving the pack rape of a refugee woman, a prominent cleric sitting on the board of a major overseas aid agency remarked, 'I hope she enjoyed it!'. Horrifying as this statement is, this man expressed a very commonly held view of rape and sexual abuse. At the time, apart from a few Amnesty International reports, little written material was available on the rape and sexual torture of women. Only later significant literature became available to support the evidence I had collected. In fact, several times I was accused of 'overreacting'.

Rape as Torture
Much of the rape and sexual torture of refugee women is planned and systematic. In camps, it is institutionalised and a method of control. These acts are undertaken with relative impunity. The women told stories of how, in situations of conflict, rape was used as a strategy to shame communities, destroy social fabric, and extract information. They were forcibly impregnated to destroy ethnic purity. From Latin America came stories of genital mutilation with electric prods, with broken glass, and the use of trained dogs. Some of the stories we collected were just too terrible to tell. In some cases, women and girls died from the injuries they sustained. In a perverse way, the stories explained, but could never excuse, why the term 'only rape' was used, because some refugee women experience such violent sexual torture. The absolute horror was in the systematised and deliberate way that rape was perpetrated against women. This was not just an opportunistic phenomenon when men found themselves in a position of power over vulnerable women. Simply changing policy language and distributing information brochures was futile (Compiled field notes, Pittaway, Hong Kong, Thailand, Malaysia and Indonesia, 1989 and 1990 and Australia 1991).

As refugee policy is strongly linked to international human rights conventions, and as these were such blatant violations of human rights, we hoped that a solution might be found by recourse to these instruments. This demonstrated how little we understood the human rights framework! I undertook a major literature survey in order to reference the WaR policy directly to the appropriate human rights conventions. However, the relevant human rights instruments did not adequately recognise or address the issues of the rape, torture and trauma of refugee women. Not only did these instruments not provide a solution, they were also part of the problem.

> The issue was not only invisible in Australian policy, but internationally. Rape in conflict situations, which includes systematic rape and premeditated sexual torture, was not considered as a crime against humanity, a war crime or grounds for refugee status. The rape of a woman was not considered an act of torture.

The literature survey identified the gender silence inherent in human rights instruments. Human rights instruments conventionally address the 'public sphere' of community and political life and overlook the 'private sphere' of the domestic arena. This is the domain of most women and that to which rape and sexual abuse had been relegated. Because of these anomalies, the human rights instruments did not specifically recognise the rape and sexual violence experienced by women as torture. We were hopeful when we discovered that the preamble to the Torture Convention acknowledged rape as torture, but the operating paragraphs in the directions to the UN Special Rapporteur on Torture refer to torture *and* rape. These semantics, seemingly small changes in language, provided the basis for the dismissal of rape as torture. Judges declined to accept it as the grounds for refugee status because '[it] is the common experience of women everywhere' (Castel, 1992, p. 39). Many similar cases were documented in Haiti, Kashmir, Tibet, Peru, countries in the Horn of Africa and the former Yugoslavia. It was a universal problem (Human Rights, 1995). It is best summed up by Williams:

'*Perhaps the most extreme example of the gender blindness*[1] *of the Geneva Convention is the situation of refugee women who have been victims of organized rape by soldiers or government officials. They have faced difficulties because the authorities tend to view rape as an apolitical, 'personalized' act of random violence against an individual rather than a component of a political strategy against persecuted groups. People tend to misunderstand the substantive function of rape; they see it as a sexual act, even think it's acceptable, a kind of morale booster or booty in war. They fail to see it is an act that is about power and about terror*' (1993, cited in Bhabha, 1993, p. 245).

[1] We have endeavored to use anti-oppressive language wherever possible, preferring the term gender-silent over gender-blindness; however, we present quotes from earlier years in their original form.

From the Local to the Global

The understanding gained from the re-evaluation of the research findings and the literature survey was a turning point for ANCCORW. We realised that if the abuse of refugee women and girls was to be addressed, three things had to happen. One was that the rape and sexual abuse of refugee women had to be recognised as a war crime and a form of torture. Two was that this must be recognised as persecution and thus as grounds for refugee status. Three was that it had to be reflected in international law and policy. Without this happening, programs designed to address the experience of refugee women would constantly fail. This lack of acknowledgement not only explained the failure of the WaR Program, it also clarified why the rape and sexual abuse of refugee women had not been named and addressed. Other groups around the world were also coming to this conclusion. Ann Brazeau, the head of the Women's Bureau, UNHCR, had requested that ANCCORW work through the Commission on the Status of Women (CSW) to advocate at the Fourth World Conference on Women to be held in Beijing in 1995. She was keen to involve other UN agencies in the fight for equity and social justice for refugee women. She argued that getting issues to the UN General Assembly (UNGA) through a number of UN agencies would be an effective strategy. An International Caucus on Refugee Women, based in Geneva, led by Beth Ferris from the World Council of Churches, and a Human Rights Caucus organised in New York, led by Charlotte Bunch, were also working to achieve this aim and Velvy Holden and I linked up with these groups. From this point onwards, our national and international activities and advocacy were inextricably linked.

A Racially Motivated Response

While we were part of a network of committed human rights and refugee activists, it was still difficult to gain acknowledgement of rape in conflict in the wider community and the media. In the 1980s and early 1990s, the majority of refugee women came from developing countries and regions including Indo-China, Iran, Iraq, the Horn of Africa, and Central and Southern America. They were not Caucasian, held lesser status than the major decision-makers and power brokers in the world and were viewed as subjects of pity and charity rather than people with equal rights. In 1990, however, war broke out in the former Yugoslavia, and because this region is a part of Europe and accessible to Western media, it received much more coverage than wars in other parts of the world. The sexual abuse of women, the 'ethnic cleansing' through the killing of men and impregnation of women and the 'rape camps' that occurred in the three countries involved were nightly television news, inciting international outrage.

This anger was intensified because the women were Caucasian and the villages and towns were those of a developed country. The average person in the Western world could identify with the women and their experiences in a way that had not happened before. Similar treatment of refugee women from developing countries was well documented and reported in the past, but never received the same level of response. AUSTCARE, a humanitarian aid agency, and UNHCR Australia started a major campaign to send 'comfort packages' to women in the former Yugoslavia. Juliana Nkrumah, an ANCCORW board member wryly commented, '*there have been African women experiencing what those women are experiencing for many years. Do they think that we don't bleed?*'. Despite the world spotlight, the majority of women raped and sexually abused in the conflict in the former Yugoslavia were not accorded refugee status. Those who entered Australia as the result of these atrocities came on Special Humanitarian Visas. The lack of recognition of their experience as a crime, sufficient to warrant full international protection denied the gravity of the experience they had suffered. *Racial privilege did not overcome gender discrimination.*

Acceptance of the magnitude of the sexual abuse and rape of women was and is extremely confronting. The world had to digest the fact that it was not just a handful of bad men perpetrating these atrocities. In the same way that it was difficult to accept that it was Caucasian women being raped, it was equally painful to realise that it was Caucasian men doing it (Mladjenovic, 1993). This challenged many men, who felt a sense of collective blame. It challenged women who had to accept the fact that many men, placed in situations of power, will do this. It was a strong painful realisation about gender relations and a difficult concept for many to contemplate. There was a backlash against the more militant feminists advocating for this cause, who disrupted marches commemorating war and protested next to statues of the military. A partial solution to the dilemma was proposed by some human rights groups and the media, and taken up internationally. They introduced a new concept into the discourse: rape in a conflict situation was identified as a 'weapon of war'. This new language was palatable to the public because it took the male agency out of rape, and it was a powerful force for gaining wide public support for refugee women's suffering. It helped to break the silence, and the rape and abuse of refugee women began to be more openly acknowledged at the UN and by governments around the world.

Rape as a Strategy, Not a Weapon: An Example of the Power of Discourse

Rape is an intentional act, an act of violence, the majority of the time perpetrated by a man against a woman. To commit the act of rape, a man has to be both psychologically and physiologically prepared to do it. The major weapon is his own body. He cannot disassociate himself from the act. It cannot be done long distance. It is not done in self-defence. To claim that rape is a weapon of war is to give the action a dignity and an acceptance which it must not have. Rape is a violent crime against an individual and a direct assault on the sexual identity of the woman concerned. To accept it as a 'weapon of war' is, in many ways, to sanction the action, to accept it as an inevitable part of the horror of the war experience. By conceptualising women as victims of a weapon, akin to other weapons of war such as landmines, it is not necessary to examine the psychosocial implications, and therefore easier to accept. The wide acceptance of the phrase 'weapon of war', which allowed people to distance themselves from the action and focus on the figurative weapon, was used by many groups lobbying advocating to have rape classified as a war crime. The result was the further objectification of refugee women as victims of circumstances beyond their control and in need of protection rather than empowerment. However, the unauthorised use of a weapon is a crime, especially against civilians. Some human rights networks worked hard to play down the implications of the term 'rape as a weapon' in both the media and public documents. We constantly reinforced the concept of rape as a criminal act and an act of torture. Whenever possible, we challenged the notion of rape as a *weapon* of war. As a result, the UN subsequently adopted the phrase 'rape as a strategy of war'. While an improvement, it still failed to acknowledge it as a crime.

The Uneasy Coming Together of Advocacy at Domestic and International Levels

In Australia, resistance to advocacy about refugee woman as survivors of rape came from some migrant women's rights advocates, as this was seen to shame both the woman and her community. Based on their struggle to challenge the prevailing discourse of racism, some feared that negative stereotypes of refugee women would feed into this discourse. Little resistance came from refugee women, and most were relieved to have the issues out in the open. Care was taken to present an accurate picture of women as *survivors* of rape, the guardians of family and culture who bring great strength and experience to a new country, not as helpless victims who require protection and care. ANCCORW had by now established a number of refugee women's groups in Sydney, who spoke out about rape in conflict and their needs in resettlement. They held functions and developed advocacy strategies for international and domestic policy. During 1992, ANCCORW committees of refugee women were also formed in Melbourne, Adelaide, Darwin, Perth and Brisbane, and I travelled extensively to these centres to work with refugee women and to identify their issues of concern. Whenever possible, refugee women attended meetings with government and service providers to represent themselves, but lack of funding to do this was always a problem.

Some of the large charities and church groups, who were part of the refugee policy community, were also upset with ANCCORW because they had been criticised by refugee women in the ANCCORW report on resettlement services. This was tempered by the fact that the status ANCCORW had gained through its international activities was giving us access to government. As a result, issues for which the actors in the refugee network were fighting had gained substantial public support. Shared concerns about the direction of service provision saw the eventual coming together of the various players in a somewhat uneasy, but extremely productive, alliance. A committee called the Refugee Resettlement Working Group (RRWG) was formed in August 1993, following a meeting convened by the director of a Migrant Resource Centre (MRC) in western Sydney. The RRWG committee consisted of more than 20 representatives from major national and state charities, religious groups, community organisations, refugee groups and relevant government departments. I was appointed chair of the committee in January 1994. The aim was to develop a coordinated approach to the issue of refugee resettlement and the WaR visa program in Australia, and to present it to the Government for consideration. This was achieved and it was presented to the Government at a high-profile conference and received very positive media coverage. For nearly a decade, it informed important changes in resettlement services. However, despite these gains, we were unable to address the problems with the WaR Program overseas.

ANCCORW had by this time received substantial government funding to run labour market programs for refugee women, and in 1994, it was decided that it would be preferable for ANCCORW to be constituted in its own right. The new body was renamed the Australian National Committee on Refugee Women (ANCORW) Cooperative.

Onwards to Beijing

In 1992, excitement was building in the Australian women's movement and overseas about the upcoming UN Fourth World Conference on Women, to be held in Beijing in 1995. A coalition of women's NGOs in Australia, the Coalition of Participating Organisations of Women (CAPOW!), was founded at the Women's Electoral Lobby (WEL) national conference. Its role was to coordinate activities linked to the conference and to facilitate submissions to the Federal Government and UN bodies involved. The ANCORW Board made the strategic decision to actively seek ways in which to work more effectively with both refugee and feminist policy networks in Australia and joined CAPOW! In November 1993, I was elected co-chair. By this stage, we had learned that the process of engagement with a wide range of like-minded people provided numerous opportunities for advocacy on refugee women's issues.

Planning for the Beijing Conference began in earnest in March 1993, at CSW 37. Member states had been asked for input into the first draft of the Beijing Platform for Action (BPFA), the Outcomes Document to be produced from the conference (UN, 1995). The BPFA aimed to provide a blueprint for achieving gender equality across the world by setting strategic objectives and actions to address 12 areas of concern. These were: women and poverty, education and training of women, women and health, violence against women, women and armed conflict, women and the economy, women in power and decision-making, institutional mechanisms for the advancement of women, human rights of women, women and the media, women and the environment, and the girl-child. ANCORW wanted the rights of refugee women included in all areas of concern, but our main foci were those of women and armed conflict and women's human rights. Following discussions with our international allies, our major aim was to have rape in conflict recognised as a war crime in the BPFA. This would provide a metaphoric hook upon which to hang the much needed law and policy. While the BPFA would not have the status of international law, it would carry considerable 'moral imperative' for governments to undertake actions based on a document they had publicly endorsed and signed. With our growing understanding of the UN process, we knew that we needed to be present not just at the Conference, but at all the meetings leading up to it. Despite the difficulty of a small NGO raising the funds to send representatives to participate in the process, the ANCORW Board decided that it was important that we be there and funding from a wide range of sources enabled it to happen.

Building Momentum: The Preparatory Meetings for the Beijing Conference

From 1993, until the Beijing Conference in September 1995, a series of international meetings took place. Each regional office of the UN Economic and Social Council (ECOSOC) convened a high-level ministerial meeting. There was so much interest in the Conference; it was decided to limit the number of NGOs who could participate in the formal conference and hold a parallel NGO Forum of related events for others who wanted to attend. An international NGO committee was convened and a secretariat established at CSW to organise the Forum. The five regions of the UN all held NGO preparatory forums. The outcome documents of the high-level ministerial meetings and the NGO regional forums were all taken into consideration as part of the official preparation of the Conference and Forum, and in drafting the BPFA. During this period, ANCORW gained accreditation to attend ECOSOC meetings in its own right. I represented ANCORW at the ministerial and NGO meetings in the Asia-Pacific and planning meetings in New York. Remembering my feeling of panic and lack of preparedness at my own first UN meetings, a training package was developed, using the work of ANCORW with refugee women as an example of how women's groups could mobilise for the Beijing process. The benefit to ANCORW was that it provided a forum for the dissemination of information about refugee issues and gained wide support for our advocacy, both in Australia and on the international stage.

The Asia–Pacific NGO Regional Planning Meeting, Manila, November 1993

The first regional planning meeting was held in Manila, November 1993, and attracted 600 participants from the Asia-Pacific region. It was organised by a committee of NGOs appointed by CSW, as part of an international NGO working group, and I was elected to the working group representing CAPOW! Issue-specific caucuses were formed to formulate strategies to influence the twelve areas of the BPFA. ANCORW sponsored a number of refugee women to attend this meeting and they worked in the refugee and human rights caucuses. This was an excellent opportunity to gain support for our position on refugees from NGOs across the region. NGOs do not only lobby at the UN, they also lobby their own governments about the positions they will take at meetings, and they influence other NGOs.

While in Manila, I worked with and was invited to join the Asian Women's Human Rights Council (AWHRC), a group of women from 16 countries in Southeast Asia. They worked together on women's human rights issues in a loose alliance, led by the indomitable Corrine Kumar, a human rights activist from India. Their major activity was running 'Human Rights Courts of Justice' on women's issues in different countries, often attended by several thousand people, and including renowned human rights lawyers and activists as 'judges'. I found myself in the privileged position of being the only Council member who had not been imprisoned because of human rights activism. It was a humbling space to be in and provided me with a crash course on the meaning of human rights to women from the global south. By the end of the Manila meeting, a regional NGO position on refugee women was formulated and unanimously accepted as a position to take to the first regional ministerial meeting in preparation for the Beijing Conference to be held in Jakarta, June 1994.

Back in Australia, February 1994

ANCORW's expanding international networks also proved crucial in improving our relationship with the Australian Government. Information about the Beijing meetings was often very slow to reach the Federal Government Office for the Status of Women (OSW). Through our international networks, ANCORW often received information on the meetings considerably before OSW and we would pass this information to them until they were able to establish direct links with the conference organisers. This helped to establish a strong working relationship between ANCORW and OSW staff. From a rather strained relationship at the outset, a closer cooperation began to develop. Many of the bureaucrats involved had little or no experience in UN meetings, and ANCORW's considerable knowledge of process and procedure was useful to some government representatives. In February 1994, members of the ANCORW Board and other lobby groups were invited to Canberra to discuss the various issues of concern to the refugee sector which could be addressed at the international level. They were unanimous in requesting the government to support the inclusion of rape in conflict situations as a war crime. The government agreed.

Financial support was provided through OSW to train and support Australian women to participate in the Beijing Conference. CAPOW! and ANCORW received funding, and a person was employed to prepare information about the Conference for refugee and migrant women in community languages, and to conduct information sessions across Australia. In response to strong lobbying from national women's groups, OSW agreed to include NGO members on government delegations for future OSW and Beijing-related meetings. Following discussion with the ANCORW Board, I did not nominate for this position. It was felt that NGOs on government delegations were sometimes restricted in their lobbying and unlikely to exert major influence on the delegation from the inside. In UN meetings, NGOs have greater power as 'free agents', but is invaluable to have an NGO in that position to facilitate communication between the two groups.

The Regional Expert Group Meeting, Bangkok February 1994

While the first meeting in Manila was about the NGO Forum, this was about compiling expert evidence to inform the draft BPFA. This would be presented at the ECOSOC Asia–Pacific ministerial regional meeting held in Jakarta in September 1994. At this meeting I was invited to join a second women's network, the Asia-Pacific Women's Watch (APWW), the NGO coordinating body for the region. APWW divided the Asia–Pacific into five sub-regions, each of which had three elected representatives and I was elected as the Australian representative and the refugee focal point.

While more conservative than AWHRC, APWW had excellent connections to governments in their own countries and amazing knowledge and experience of working inside the UN. The following year, I was appointed as co-chair and quite often presented the public face of the group. When questioned as to why APWW had a white woman from Australia as a co-chair, the other co-chair explained that because of the sensitive nature of gender issues in the Asia-Pacific, and the patriarchal nature of some of the governments, it was sometimes not safe for them to speak out publicly. Having a brash Australian speaking meant that I could say publicly (but still diplomatically!) things that they could not. If they were challenged about what I said, the agreed line was 'oh that loud Australian, she just says what she wants!' It was a sad subversion of white privilege, and in a strange—but uncomfortable—way, it worked. The work and support of APWW were critical in raising refugee issues in the region. While in Bangkok, I also attended a regional meeting of UNHCR to argue for rape in a conflict situation to be recognised as a war crime and reinforced the need to acknowledge gender as grounds for persecution in refugee law. This was an important part of the strategy of using the Beijing process to gain support for changes proposed in UNHCR policy and practice.

The First International Planning Meeting for the NGO Forum, New York, March 1994

This meeting happened just before CSW 38. I attended as an observer/lobbyist representing ANCORW, AWHRC and APWW's positions on refugee women. We worked closely with the international human rights group, who focused on refugee women as one of their key concerns. This group had many prominent human rights lawyers as members. Here, I gained a much deeper understanding of the intricacies of working within the UN lobbying system and the importance of using tried and tested legal language, preferably from documents, laws and conventions already agreed and signed off by governments. At times, plagiarism can be a positive! Bringing together the work from the five regional meetings, strong international caucuses were formed to plan lobbying strategies leading up to, and including, the NGO Forum and the Beijing Conference. The importance of attending the UN regional ministerial meetings was clear. Reports from these meetings carry significant weight, and governments have a 'moral imperative' to implement agreements in them once they have signed them. They are seen to represent the common good of a particular geographic region. History has shown that governments are more likely to sign off on issues at international meetings if they have debated them previously at the regional level.

The aim of the refugee caucus was to have the issues of rape and sexual abuse of refugee women and gender-related persecution acknowledged in international law. The head of the Women's Bureau, UNHCR, urged that we stay with these issues and not be side-tracked into a focus on service provision in refugee camps and the protection of refugee women. She needed the support of NGOs and governments to assist her work within UNHCR. The mainly male hierarchy supported protection and improved service provision as the answer, rather than acknowledgement of the endemic sexual abuse. She argued that if we addressed the root cause, then improved service provision would follow. There was a strong contingent of NGOs at this meeting and substantial support was gained from women's groups which previously had little to do with refugee issues, such as the YWCA, the Girl Guides, Women with a Disability and the Soroptimists. In turn, members of the refugee caucus supported the positions of these other groups. The method of Australian NGOs working cooperatively at international meetings continued and was refined.

UN Economic and Social Council Asia-Pacific Jakarta, September 1994

Following quickly on CSW 38, the regional ministerial meeting for the Beijing Conference was held in Jakarta. The purpose of this meeting was to produce a regional governmental document, the Jakarta Declaration, outlining commitments to improve the status of women in this region and to provide agreed regional input into the BPFA. NGOs attended as observers, and our role was to lobby member states to include issues of concern in the BPFA. Thampuying Sumalee, the Thai president of APWW, had produced a fantastic lobby document, 'The Little Yellow Book', a comprehensive compendium of the issues APWW members wanted included in the BPFA, with suggested language. Because of her position as a member of the Thai Royal Family, she was able to organise a formal presentation of the booklet to all government delegates at the meeting, and it was very well received!

Prior to the ministerial meeting, AWHRC held a seven-day intensive workshop on human rights lobbying, run by two prominent international human rights lawyers. As part of the workshops, correct legal and UN language was identified for each of AWHRC's advocacy priorities, including refugee issues. At the ministerial conference, we were able to work strategically with the majority of governments present, not only talking about the issues in the document, but also providing legally correct and acceptable language for governments to use and support. The strategy was extremely successful. Wearing the triple hats of APWW, AWHRC and ANCORW, I was able to make a presentation on refugee women's human rights to the full plenary session on the first day of the conference, an excellent time to present. Sadly, we were still a long way from insisting that it be refugee women who should attend and speak at these meetings!

The Australian Government delegation was impressed with the quality of the work done by the regional NGO working group and incorporated much of the language prepared by them into the Jakarta Declaration. On several occasions, I was invited to work with the government delegation and with representatives from other governments as issues were discussed and positions reached. A system of daily briefings was instigated between the Australian delegation and the NGOs to ensure a flow of information and strategies between the two groups. On two occasions, the government recruited NGOs to negotiate unofficially with other governments over issues, when diplomatically they could not make a direct approach. One Pacific Island nation recruited an NGO representative to assist in drafting their entire response to the meeting, as they were not familiar with the language process. The Jakarta Declaration contained a commitment recognising gender as grounds for persecution, rape as a war crime and support for improved recognition of the needs of refugee women (UN, 1994).

The Politics of Policy; The Theatre of Policy

On the third day of the meeting, at 2 am, consensus had still not been reached on the document, which was to be launched eight hours later. A major sticking point was that of rape as a war crime, with one powerful nation blocking the acceptance of the language (it was still in square brackets!). Everyone was exhausted, and the chair was so frustrated that he asked the male representative of the country blocking the language to

> go into the corridor with the Australian Government representatives who were sponsoring it, to come to some agreement. They invited me to join them. The man asked for an explanation of why it was so important and I explained our position. He listened, and I could see that he understood its importance. He then said 'Look, we cannot be seen to be backing down on our position. I am not going back into the meeting, I am going to bed. I want you to go back in and tell them that is what I have done, and ask for the square brackets to be removed. Tomorrow morning, I will take the flag (ask to speak) right away and say how upset I am that the square brackets have been removed. I will then say that, as a gesture of good will from our government, we have decided that the language can remain. Well done and goodnight'. With that, he disappeared down the corridor. The Australian representative did as had been suggested, and right on script the next morning, he did what he had said he would do. The language stayed in the document and was signed off by all governments.

Following the meeting, the Australian Government delegation was public in its praise of the NGOs who had worked together as an advocacy team, acknowledging our expertise in both the subject matter and the process. CAPOW! was delighted with the lobbying success. The most important gain for the refugee domain was that the Australian Government decided to act on the Jakarta Declaration prior to going to Beijing. Significant changes were made to policy addressing the entrance of refugee women to Australia. It was suggested by some NGO representatives from the Refugee Resettlement Working Group (RRWG) that this was done to enable the Australian Government to present a good profile at the Beijing conference. For us, positive change was welcome, regardless of motivation. My experience at CSW held me in good stead when I attended UN meetings. The same government departments: DFAT; OSW; and DILGEA, were involved, and I was respected, if not always liked, by actors in all three, who made quasi-humorous comments such as, '*oh yes, have we got a file on you* (an Australian Intelligence Security organization file)!' The informal power I had at times was a direct challenge to their formal power. Despite this, a solid body of support developed with bureaucrats who were supportive of the position ANCORW was taking and who were happy to facilitate access to key players and, at times, to leak information, since they trusted me not to disclose my sources.

Pride Comes Before a Fall!

Despite this success, in September 1994, ANCORW, along with several other advocacy groups, learned that its funding would not be renewed. The *sacred* reasons given by the Government for this were that advocacy groups were no longer necessary, because the refugees' needs had now been identified and that services were provided to address them. The *profane* reason was that the Minister had vowed to crush the power base of the refugee lobby. Detailed information of the Minister's position was leaked to me by an extremely senior government official, who said, 'use this information however you want to, but if I go down because of what you do, I will take you with me'. Despite a coordinated national lobbying campaign, including intercessions by the head of DILGEA NSW on behalf of ANCORW, the Minister remained obdurate and began a phasing out of government funding to advocacy groups (excluding some national peak bodies). An anomalous situation then occurred, when the department which had defunded ANCORW because of its advocacy role appointed me to a federal resettlement advisory committee in October 1994. Sufficient funds were available to employ me until March 1995, but my mortgage loomed large over my head, and I accepted alternative employment as a senior policy analyst with the NSW State Government. I continued as chair of ANCORW and undertook the advocacy work as a volunteer. Good advocacy can be both rewarded and punished.

Regional NGO Working Group Meeting, Bangkok, November 1994

A final meeting of the Regional NGO Working Group was held in Bangkok, in November 1994, the last I attended as an employee of ANCORW. The meeting officially coordinated the roles people would take in Beijing, including in the regional events at the NGO Forum and the Conference itself. Most importantly, it synthesised input from each of the regional NGO caucuses into the relevant sections of the draft BPFA. Strong language, referenced to international law and human rights conventions, supported each of the Working Group's recommendations.

Final Preparations for Beijing

Fortunately, I was able to continue working on the Beijing process. Between March and August 1995, representatives of women's groups who gained accreditation to the Beijing conference were trained by ANCORW on legal language and UN processes. As well as the work of CAPOW!, the Government supported another very successful advocacy group, the Australian Council of Women, who focused on including women from across Australia in the Beijing Process. It was led by a highly respected business woman and feminist, Wendy McCarthy. Unfortunately, just the day before we were due to leave, Wendy could not travel. In my capacity as Chair of CAPOW! I was asked to fill her place. This made my time in Beijing extremely busy.

The Fourth World Conference on Women, Beijing, September 1995—The NGO Forum

The major events of the four-year Beijing process were the formal high-level government meeting, the Fourth World Conference on Women and the NGO Forum. Over 30,000 women from around the world registered to attend the Forum, but most were not accredited to attend the formal Conference. The Chinese Government became worried that the tightly controlled facilities in Beijing would not cope with the volume of people and feared that political protests might take place and disrupt the actual conference. The Forum was moved to Huairou, a semi-rural town about 60 kms from Beijing, where the majority of events took place in a huge tent city. Most participants were bussed there from hotels in Beijing, and moving around was a challenge. However, the government designated one lane on the highway from Beijing to Huairou for people travelling between the two events. To the amusement of participants, it was sign-posted 'The Fast Women's Lane'.

ANCORW members ran workshops at the Forum and took part in many activities in the 'Refugee Tent'. At a major forum event, at which representatives of the five UN regions presented reports on the state of women in their regions, I was honoured to present the Asia-Pacific Report, joining the USA/European representative, Hillary Clinton, on the stage. The report was carefully compiled by a group from APWW, CAPOW! and other leading women's organisations, and, contrary to conventional wisdom about reports compiled by committees, it came together very well, with a succinct summary of the issues we wished to have covered, including, of course, rape as a war crime. While there was little formal communication between the two events, those women with dual registration came and went and acted as vectors, informing their governments about key issues coming from the Forum. Westpac Bank funded a meeting for Australian NGOs, in a large hotel in Beijing every evening of the conference. This allowed Australian women who were not accredited to attend the UN Conference to be informed of the progress and to pass issues raised at the Forum to delegates at the conference. The work of coordination was shared between a group of fantastic women, in particular, feminist activists and excellent organisers, Trish Szirom, who single-handedly took on some very intrusive interventions by the local security forces, and Libby Darlison.

Unfortunately, the weather was not kind, but this did not spoil the atmosphere or the outcomes. It was a joyful and inspirational theatre of events. It provided a venue for information exchange, consciousness-raising and education for the many thousands of women who attended it. However, it was also rife with politics, sometimes between women's groups with very different ideological positions, sometimes on a racial basis and often between government representatives, who also attended the meetings.

The Dirty Politics

One of the women who had come as part of the Australian delegation was a refugee from Tibet who had Australian citizenship. She was closely monitored by Chinese security and often threatened as she moved around. Things came to a head at the Forum, at an event when the extremely popular Australian Minister for Women, Carmen Lawrence, came from Beijing to run a session on women's rights. It was in an open space and a large crowd gathered. At some stage, it became obvious that the Tibetan woman was being seriously harassed and threatened. A large number of armed men were massing around her. A message was sent to the Minister, who did not blink an eye. After a short word to her aide, she called out that she wanted to honour some of the Australian refugees in the group and called them to where she was speaking. They of course included the woman from Tibet. As soon as they joined her, the Minister invited the crowd to form circles around them and join hands in solidarity with the women while she spoke. Quickly, a tight cordon of women encircled the refugees and the minister. Within a very short time, a large car was seen to be slowly pushing its way through the crowd across the muddy field. It was the diplomatic car of the Australian Embassy, complete with flag and covered by diplomatic immunity. It slowly nosed through the circle to the Minister. Out jumped the Ambassador, who ushered the Tibetan woman into the car and drove off. Without blinking an eye, the Minister carried on with her presentation. That night, at a reception hosted by the Australian Embassy for the 60 Australians in Beijing with much champagne was consumed. The Minister and the Ambassador were the heroes of the hour. The Tibetan woman continues her advocacy work. This event highlighted the reasons why people become refugees and the impact this has on their lives. It made the women from Australia even more determined in their advocacy.

The Inter-Governmental Conference

The Beijing Conference was held amid very strict security. Not only were the venue and major hotels cordoned off, but whole suburbs were also out of bounds. The conference was the culmination of all the lobbying and processes that had occurred to date. It built on political agreements reached at the three previous World Conferences on Women and consolidated five decades of legal advances aimed at securing the equality of women with men in law and in practice. More than 17,000 participants attended, including 6,000 government delegates, along with more than 4,000 accredited NGO representatives, a host of international civil servants and around 4,000 media representatives.

The draft BPFA had been continually amended and refined during the four-year preparatory process. It contained input from government reports, NGO reports and expert group meetings from the five UN regions of the world, and three meetings of the CSW. Final amendments at the conference were debated through 'Square Brackets' and 'Language' sessions. The task of the Beijing Conference was resolution of the language still in square brackets. At this stage, although there was ongoing resistance to addressing rape as a war crime in the BPFA, the opportunity was still there. Fundraising had ensured that a team of six women would represent ANCORW at the formal Conference, and we were well trained and coordinated. We had prepared 'language' and had the wording checked by an international lawyer. Even at this late stage of finalising the BPFA, many governments were pleased to have 'language' presented to them.

Shock Value as an Advocacy Strategy

As NGOs, we needed to present evidence to member states to support the changes we requested. A brilliant international lawyer from the USA, Donna Sullivan, presented a case study that gave our case huge momentum. In a classic case from El Salvador, a man was tied to a chair and forced at gunpoint to watch his common law wife being raped by nine soldiers. In determining the case for refugee status, he was deemed to have been tortured and granted refugee status. His partner was not, as she had only been raped. Because they were not legally married, his refugee status did not cover her and she was denied international protection. The sheer hypocrisy of this legal position went a long way to persuading member states to support the language of rape as a war crime.

The ANCORW members were active and effective, working as a lobby group, systematically targeting other governments and influential NGOs. The range of different languages spoken in the group assisted in this process, and networks established at previous meetings were invaluable. On several occasions, we were invited by members of the Australian Government to participate in drafting language for the final document. Issues concerning the successful resettlement of refugee women, and in particular, refugee women at risk, were constantly raised with government delegates who spoke very strongly and positively on refugee issues. Discussions continued to the very last moment, and on the final days, the sessions ran into the night. At times, it was more like a cattle market than a conference on women's rights, with governments trading concessions and bargaining for agreements, mainly on ideological grounds. For example, after lengthy discussion, the right of women and girls to inherit their husband's and or father's property was included on the condition that all reference to lesbians and sexuality was removed from the draft document. It was into this political quagmire that we took our last fight. Eventually, in the early hours of the morning on September 19, the BPFA was adopted unanimously by 189 countries, providing a strong agenda for women's empowerment. It is considered the key global policy document on gender equality. Much of the wording prepared by the Refugee Committee of the Human Rights Caucus was incorporated into the BPFA. In all, fourteen paragraphs of the document relate to refugee women and girls.

We Won!
Paragraph 145D of the BPFA states: That Government shall*: Reaffirm that rape in the conduct of armed conflict constitutes a war crime and under certain circumstances it constitutes a crime against humanity and an act of genocide* as defined in the Convention of the Prevention and Punishment of the Crime of Genocide: 27, take all measures required for the protection of women and children from such acts and strengthen mechanisms to investigate and punish all those responsible and bring the perpetrators to justice (UN, 1995).

The powerful advocacy by the various lobby groups, with strong support from the Australian Government delegation and other member states, had paid off. The ANCORW team had been an effective part of this collaborative international process. Not only had we collectively achieved our aim, but the recognition that, at times, rape was a crime against humanity and an act of genocide made it even stronger. The specific needs of refugee women were also included in other sections of the document. The solid lobbying and the irrefutable evidence presented had worked!

The Australian Government signed the BPFA without reservation. However, we knew from experience that words on a page do not necessarily translate into action. Even while celebrating we were aware that this was not the end, but the beginning of the fight for implementation. One strategy we were already using to push for domestic implementation of the BPFA in Australia after the Beijing Conference was the use of the media. Working with ANCORW, the Special Broadcasting Service (SBS) Television station produced a 30-min current affairs program on the issue of rape as a war crime. The first 25 min was compiled prior to the Conference and comprised conflict scenes from around the world, interspersed with commentary about the work of ANCORW. The last five minutes were recorded live by telephone interview with me from Beijing, announcing the inclusion of rape as a war crime in the BPFA. The program was screened the night that this happened. It received excellent coverage and raised awareness in Australia about the Government's international obligations.

Theory and Discussion

The Critical Structuralist Paradigm

> It is not the business of hegemony to remove powerlessness or redistribute either material or cultural advantages, but to rearrange perceptions in such a way as to secure consent. (Bertelson, 1986, p. 18)

The academic learning in this stage of the story was challenging. The critical paradigm provided meaning to the very confusing events in our journey and the complex ideological tensions between different actors. I love the notion, inherent in critical analysis, that the truth is but one person's version of reality: this helped me stay sane when faced with a multiplicity of 'truths' in the policy process. The concept of 'ideological hegemony' was also invaluable. This recognises that as social actors, we embody unquestioned assumptions about things like gender and race, which reflect the patriarchal and other dominant ideologies which shape society. These assumptions are so ingrained in us through socialisation that they are often unrecognised, even by activists, and thus, they 'permeate the very hegemonic order they intend to oppose' (Bertelson, 1986, p. 18). Therefore, to understand how 'policy is made', we must understand the dominant ideologies and how they impact on all the actors in the arena, including activists, advocates and the recipients of the policy outcomes. The critical paradigm is a broad church, and I focused on those parts which shed most light on the questions we were seeking to answer.

Feminist Analysis and the Policy Process

Feminist analysis of policy also provided new insights, which challenged the gendered assumptions we were encountering, and the gender bias embodied in international law. Feminist theory is broad and covers a wide range of ideological viewpoints, but common to all feminist theory is the principle of equality of women and men and the belief that this will only be achieved by challenging gender-based injustice wherever it is embodied in patriarchal social structures. Many of the gender inequalities which feminism seeks to eradicate are institutionalised and reflected in policies and laws which grant rights and privileges to men, but not to women, such as the El Salvador case, where a man forced to watch his partner raped was granted refugee status, but the raped woman was not. These inequalities are based on an historical invisibility of women in social life and institutions and a false dichotomy of the differences between men and women based on women's reproductive capacity. These differences are both enshrined in and reproduced by patriarchal ideology which informs much of our current legislation and policy. Writers such as Fraser (1989) and Newman (1995) have explored the myth of 'gender neutrality' in policy and legislation. This has highlighted the gender blindness in all stages of the policy process, including problem definition, needs analysis, policy formulation, implementation and evaluation. They argue that if a feminist analysis is not used at each stage of the policy process, the outcome is most likely to reinforce the dominant ideology and accepted 'truths' about women's place in the social structure.

Feminist analysis explained the acceptance of rape as 'the common experience of women everywhere', and the lack of acknowledgement of rape and sexual abuse of women as torture, or grounds for refugee status. What we were doing challenged the patriarchal status quo—the power of men over women's bodies. It challenged people to look again at an act which had been accepted as a sad inevitability and acknowledge that it is in fact a criminal act.

Discourse and the Policy Process

> There is a strong argument for the active role of discourse in the creation of social reality: discourse does not reveal some hidden, pre-constituted reality, but rather provides concepts, objects and subject positions that actors use to fashion a social world. (Hardy & Phillips, 1997, p. 7)

I next turned to discourse analysis to explain the power of the term 'only rape' to shape acceptance of the sexual abuse of refugee women in policy and practice. Analysts in the critical paradigm address the unconscious use and power of discourse and evaluate the role of discourse in producing social meaning. They focus on how spoken language and written text are structured and influence the social context in which the discourse occurs. The way in which language is organised into discourses has immense power to shape the way in which people experience and behave in the world (Burman & Parker, 1993, p. 1). There are many ways in which the term is used; discourse is a dynamic process in which meanings change over time as society changes (Parker, 1992).

Feminist writers have for a long time been aware of the role of discourse in the creation of social reality, and the need for a rigorous analysis of the dynamics of discourse and how discourse can be changed (Rixecker, 1994). Feminist epistemology challenges the traditional models of knowledge creation and provides alternate 'ways of knowing', recognising that knowledge is never 'value free' and reflects the actors' interpretations of experience, including emotions. 'Only rape' is a prime example of this, revealing how patriarchal tolerance of male sexual violence against women shaped the understanding of many actors about what constitutes a refugee woman being 'at risk'. Thus, by using our knowledge of women's lives, we can understand how dominant structures and ideologies marginalise people. Truth is not absolute; it changes with time and reflects the tension of situational reality derived from situated knowledge. Our challenge was to use this understanding to change the discourse about the rape and sexual torture of refugee women.

The Challenge of Structural and Operational Change

A daunting task faced us. As we were quickly learning, there are many ideological and structural barriers to achieving gender equality and addressing the ongoing protection gaps experienced by refugee women and girls. The task suddenly got much bigger, we were not just seeking to change law and policy, and we were seeking to change the social context in which it was produced!

Structuration theory provided a useful framework to understand how social structures are formed and then, in turn, produce new structures (Giddens, 1985; Stones, 2005). The suffering of refugees is structured by the events that lead to their forced displacement and by the crucial role of UNHCR, appointed as an international arbitrator in the unregulated mass movement of people around the world. The language and discourse about refugees and gender also shape the way that refugee women and men experience the world (Kahlert, 2012). The social world in which this takes place is structured by the response of states, their interpretation of international laws and UN policies, and their willingness to comply with these. It is also shaped by those states that use their power to influence the response of other states in accepting refugees. At a local level, the social reality of refugees is structured by the response of host and resettlement governments, and local communities within host countries. This response is shaped by unquestioned assumptions about gender and other ideological forces, which impact all actors and, in turn, reinforce the social structure (Kahlert, 2012). We can see this playing out in needs identification. The way that refugees' needs are defined shapes the policy response (Fraser, 1989). To date, states, donors, UNHCR and NGOs have been in control of this interpretation, using an 'expert' and often gender-silent definition of needs. Therefore, it is critical to name refugee women in every policy, ensuring that they are equally represented in all levels of decision-making, so that they become an active part of the process of structuration.

Structure is not a barrier to changing actions, policy and law, but is essentially involved in the reproduction of these (Giddens, 1985; Kahlert, 2012). Rules and resources structure and facilitate operational responses, which change, reinforce and reproduce the social structure in a never-ending process. Thus, the determination officers in Hong Kong and Manila acted as they did because of the power of the structure in which they worked, but by doing this they reproduced the structure. They worked in a structure that failed to acknowledge rape of refugee women as a problem; therefore, they accepted this and did not identify women who had been raped as needing assistance. This is why, in order to achieve lasting change, social structures must be addressed.

Conclusion

The policy analysis paradigms provided a solid framework with which to analyse our work. Critical analysis gave us the deeper level of understanding which we needed to move forward, feminist analysis helped us to articulate the gender silence of the system in which we were working, and discourse analysis was the key to understanding the iceberg of which the words 'only rape' were just the tip. This helped us to unpack the power dynamics at work and to clearly identify the informal 'profane' power ANCORW held. My appointment to various networks and groups also increased our formal power and gave our work credibility and status. Our knowledge of the UN system and use of language gave us a level of discursive power, and we used this effectively.

These insights provided a solid framework for the next cycle of the research. Another year flew passed, following up on the outcomes of the Beijing Conference and strategising how to take the commitments made in the BPFA forward. In 1997, the planning began for the first follow-up meeting, Beijing + Five, to take place in New York in March 2000. Work was ramping up and I was extremely pleased when I was joined by Linda Bartolomei, at first in the capacity of postgraduate student and intern, and quickly to become an equal partner in the journey.

References

Bertelson, E. (1986). *The unspeakable in pursuit of the unbeatable*. University of Cape Town.

Bhabha, J. (1993). Legal problems of women refugees. *Women: A Cultural Review, 4*(3), 240–249.

Castel, J. R. (1992). Rape, sexual assault and the meaning of persecution. *International Journal of Refugee Law, 4*(1), 39–56.

Fraser, N. (1989). *Unruly practices: Power, discourse and gender in contemporary social theory*. University of Minnesota Press.

Giddens, A. (1985). *The nation-state and violence*. Polity.

Hayes, D., & Winton, S. (1991). *An evaluation of the women at risk scheme*. ANCCORW.

Human Rights Watch. (1995). *The human rights watch global report on women's human rights*. Human Rights Watch.

Kahlert, H. (2012). The simultaneity of stability and change in gender relations: Contributions from Giddens' structuration theory. *Studia Humanistyczne AGH, 11*(2), 57–68.

Mahmud, N. (1996). Crimes against honour: Women in International Refugee Law. *Journal of Refugee Studies, 9*(4), 367–382.

Mladjenovic, I. (1993, February). Separation, guilt, and identity crisis: Detailed analysis of the emergence of nationalism in former Yugoslavia. *Peace News*.

Newman, M. A. (1995). The gendered nature of Lowi's typology: Or who would guess you could find gender here? In G. Duerst-Lahti & R. M. Kelly (Eds.), *Gender, power, leadership and governance*. University of Michigan Press.

Parker, I. (1992). *Discourse dynamics*. Routledge.

Parker, I., & Burman, E. (1993). Against discursive imperialism, empiricism and constructionism: Thirty-two problems with discourse analysis. In E. Burman (Ed.), *Discourse, action and the research process*. Routledge.

Phillips, N., & Hardy, C. (1997). Managing multiple identities: Discourse, legitimacy and resources in the UK refugee system. *Organization, 4*(2), 159–185.

Pittaway, E. (2001). *Only rape!: An examination of the power of ideology and discourse in the policy process with a focus on policy pertaining to refugee women*. PhD thesis, UTS, Sydney.

Rixecker, S. S. (1994). Expanding the discursive context of policy design: A matter of feminist standpoint epistemology. *Policy Sciences, 27*, 119–142.

Stones, R. (2005). *Structuration theory*. Palgrave Macmillan.

United Nations. (1994). *The Jakarta declaration, ESCAP regional ministerial meeting in preparation for the Fourth world conference on women, Beijing*. ESCAP.

United Nations. (1995, October 27). *Beijing declaration and platform of action, adopted at the Fourth world conference on women*. United Nations. Retrieved November 10, 2021, from https://www.refworld.org/docid/3dde04324.html

CHAPTER 5

Interlogue

An intermission which makes you think about what's to come, as well as the significance of other events that are happening 'behind the scenes.' (2012)[1]

A 'Roadmap' to Part Two: A Challenging Journey with Confronting Detours

At the end of the Beijing process, we were in a difficult position, because while we were delighted to have refugee women and girls' rights enshrined in UN documents, we had begun to identify the numerous barriers which prevented the women accessing them. The next stages of the story detail the building of our understanding of these obstacles and the constant struggle to see them removed. We thought it would be useful at this stage to describe them.

[1] Expressions of Substance: Enter the Interlogue (insurrbution.blogspot.com).

E. Pittaway and L. A. Bartolomei,
Only Rape! Human Rights and Gender Equality for Refugee Women,
Sustainable Development Goals Series,
https://doi.org/10.1007/978-981-19-0916-0_5

Naming the Barriers

The Spectre of Gender Inequality

It may seem trite to say that the biggest barrier to gender equality is gender inequality, but it is! This is not only the case for refugee women and girls, it is universal. All women struggle for gender equality, and it has yet to be achieved anywhere. White women, women from developed countries and wealthy women are nearer to achieving equality than others, but still major inequalities prevail. Even in the global north, women are more likely to live in poverty than men, and all over the world, women suffer disproportionately from the impact of poverty (Badiora, 2017). Girls generally have less access to education, health services, resources and paid employment than their male counterparts, and less freedom of choice about their lives; these inequalities are entrenched in patriarchal social structures (UNFPA, 2015; Omotoso & Fajimbola, 2017). The majority of refugees live in the global south and the barriers faced by refugee women and girls often mirror the experiences of women and girls in the local host communities (Parcesepe et al., 2016). However, for many refugee women, the impact of this inequality is more damaging than for women residing in their own countries. Refugee women are often traumatised because of conflict, have often experienced rape and sexual violence, may have a child or children of rape, may have lost children in flight and are dependent on charity from international aid organisations and UN bodies just to stay alive. They usually lack access to safe shelter, legal employment or justice. Without legal rights, they are citizens of nowhere. These are some of the many discriminations which cause women to be vulnerable to rape and sexual violence. The more discriminations they suffer, the more vulnerable they are. Gender equality is a distant dream. This is a clear case of structural inequality that must be addressed at a global level before gender equality can become the norm in refugee situations. This is not to argue that refugee women should be privileged over other women and girls, but rather that, to achieve justice and adequate protection for refugees, the needs of both groups must be addressed equally (UNDP, 2016). This deep level of gender inequality is created by and also reinforces the other barriers they face.

Refugee Women as a Vulnerable 'Minority' Group

At least half of the total number of refugees in the world are women and girls. However, they are most often classified in law, policy and practice as a 'minority group', along with people with a disability, youth, children, members of LGBTIQ+ communities, the elderly and people from small ethnic groups or religions. It is important to note that women and girls also constitute 50% of all of these minority groups! However, the word 'minority' is not used here to refer to numbers. It is used in the sociological context where minority groups are those known to suffer from discrimination and subordination because of their status. They are differentiated from those who hold positions of social power; have unequal access to decision-making, infrastructure and resources; and receive differential treatment from those in power, with their personal capacities ignored. By definition, all refugees are a minority group on the international stage.

In addition, for many years, the most common descriptor of refugee women and girls was that of a 'vulnerable' group. Despite this having been challenged by feminist and refugee rights scholars, the discourse used to discuss refugee women and girls remains embedded in the cultural, religious and ideological frameworks of patriarchal societies, both in the country of origin and in receiving countries, by donors and UN bodies. These continue to view women as either passive victims or beneficiaries, not contributors. Women and girls are portrayed as either helpless and in need of protection, or mothers. They are perceived as unable to take control of their own lives and as dependent on, and subservient to, men. This can result in paternalistic programs to protect women which further entrenches gender inequality. This naming of women as a 'vulnerable' group has an emphasis on *personal* vulnerability. However, refugee women are not inherently vulnerable but rather are exposed to situations which create or exacerbate vulnerability and human rights abuses. Multiple levels of discrimination compound the difficulties they face. It is also important to note that refugee women and girls in all their diversity are not a 'special needs' group. While sharing the same basic needs for food, water, shelter, sanitation and security with men and boys, they have different gender-specific needs and have been shown to face additional difficulties in accessing common basic needs. In order to meet the standard of gender equality and inclusion referred to in refugee policy, the needs of both groups must be specifically acknowledged and equally addressed in all legal and policy provisions.

The Most Contentious of All Minority Groups - Members of LGBTIQ+ communities and Non-Binary Peoples

While always theoretically covered by the Refugee Convention, and all other human rights instruments, these groups, more than any other, have always had to fight for recognition and access to their rights. At the beginning of this story, despite the horrendous persecution, torture and murder of these groups, which continues today, they were seldom mentioned, and if they were, it was quickly dismissed. This still happens within their own ethnic communities, in local communities, and from some governments and donor groups. There is little acceptance of their needs or what they can contribute to their societies.

In many cultures, and social/ religious belief systems in both the global north and the global south, there is a strong perception that there are only two genders, male and female, and that one's sex or gender is assigned at birth. In all cases, these roles conform with traditional social constructs of masculine and feminine identity, expression and sexuality. There is little acceptance that gender is socially constructed. Refugees, as with many other people who identify as lesbian, gay, transgender, bisexual, intersex, queer or non-binary, suffer from extreme levels of discrimination. Slowly, over the years, this has been challenged by activists both within, and by those supporting refugee communities, but still there is little acceptance of their needs or what they can contribute to their societies. UNHCR has taken a very positive stance on this, including these groups in their consultations with refugee communities, in policy documents and in service provision when funding can be identified. However, this acceptance is not reflected by the majority of the governments who are members of the UNHCR Executive Committee. Time after time, in major policy documents which have to be signed off by the member states we see LGBTQI + communities deliberately excluded from mention. They are scornfully referred to as 'the alphabet people' and many governments continue to deny their existence in their countries. The exclusion of these groups in policy is often used as a trade-off for the acceptance of other contentious issues. The divide is mainly across the North South lines, but not totally, and is linked to the prevailing ideological and cultural positions of the governments concerned. Slowly, things are improving, some services are being offered, and UNHCR is reaching out to the communities to hear their concerns and to try and identify ways in which to respond more effectively. However, this is happening at a snails' pace and is one of the ongoing major human rights failures in refugee protection.

Undermining the Protection of Men?

The question is often asked: 'Why such a focus on the sexual abuse of women and girls? Men and boys also get raped and are important too!' Of course they are! No human being should suffer from rape and sexual abuse or any other form of persecution. However, there is also the indisputable fact that, while the majority of refugee women and girls endure rape and sexual abuse as part of the refugee experience, the percentage of men and boys who endure it is significantly lower, and because the rape of males is not normalised, it is more likely to be regarded as torture. Boys do not become pregnant, and their chances of marriage and a normal life are not quite as seriously compromised as those of a female. However, they still need specialised care and attention and services to help them address the trauma. We acknowledge that the needs of both groups are equally important, *but different*, and without effective policy frameworks and tools, these differences are not recognised.

Acknowledging the Achievements, Capacities, Knowledge and Strength of Refugee Women

Despite the layers of discrimination faced by refugee women and girls, they are not just passive victims. In many camps and refugee sites, women run crèches for children, arrange care for orphaned or lost children, provide safe spaces for women who have experienced rape and sexual and gender-based violence, manage scarce rations to ensure that families are fed, run small businesses to provide some meagre additional support for their families, organise basic schools and provide protection such as the building of thorn fences to protect themselves and their families (Bartolomei, 2015; Olivius, 2014). Much of this work is done without funding or external support. In the absence of men, women take on all roles in the family and community. These activities are undertaken by women who hold a wide range of formal and informal skills, and have a huge capacity. They also have a keen analysis of the problems experienced in camps and cities, and of potential solutions. However, because of their minority status, and the discourse of vulnerability, their capacities, skills and abilities often go unrecognised. Women are silenced by culture, tokenism, gender stereotypes, lack of access to education, lack of funding for targeted programs and limited access to representation at every level.

An Uncomfortable Truth—Not just a Few Bad Men

Another issue that has to be addressed is the uncomfortable truth that such endemic abuse cannot be perpetrated by just a few bad men. It implies a level of tolerance for rape and sexual abuse in society, even if only through people's inaction, and involves women as well as men. Men do not like to accept that it might be their peers perpetrating these crimes. Women deny that it could be their husbands, brothers and sons. Perpetrators employed by humanitarian aid organisations historically have just moved on to another site if found guilty of sexual misbehaviour. This implies a culpability far wider than just the perpetrators. It is the root of the many silences which have surrounded the issue and which are only now being broken.

The Weight of Privilege, Patriarchy and Racism

A major challenge to addressing the endemic rape and sexual abuse of refugee women and girls is that the hegemonic ideology oppressing *all* women must be acknowledged and addressed first. It is so much more than the protection of a vulnerable minority. It is about empowering refugee women and girls to have control over their lives and providing opportunities for them to use their own agency to be part of the solutions. We needed to understand the role of racism and the power of privilege in different cultures, and how factors such as ethnicity, disability and sexual orientation increase a woman's vulnerability. In order to try and understand this, we examined the sites where rape takes place—as part of conflict, during flight, at borders, in camps and in urban areas. We investigated the possible motives of the rapists, among whom were soldiers, police and other authority figures, peacekeepers and service providers, refugees and family members. We identified why rape is such an effective strategy of war, used as an act of torture aimed at harming the maximum number of people. We discovered the ongoing impact of this on women, children, men, families and communities, and how it shapes identities and futures. Children born of rape and their mothers can be stigmatised and excluded from their communities. We questioned the excuse of the 'weight of shame' and the defence of 'their culture' to stifle debate, which lets the international community off the hook and perpetuates impunity for perpetrators.

Rape, Rape, Rape: The Power of Words

Following Beijing, a strange thing started to happen. It was no longer acceptable to refer to 'rape', 'sexual violence' or 'sexual torture' in refugee policy and discussion. 'Rape' became reserved for special contexts, such as horrendous crimes ('the madman viciously raped the girl'), or as a political ploy to create fear and loathing of the 'other' ('the Taliban rape women'). But, as we have seen, the crime of rape is not exceptional; it happens quietly everywhere, in particular to refugee women and girls. Why can't we say the words *rape, sexual violence, sexual torture*? In policy and discourse about refugee women, these first became replaced by sexual and gender-based violence (SGBV), which next became gender-based violence (GBV), which in turn became interpreted by many actors in the refugee domain as domestic and family violence. 'Rape, sexual violence and sexual torture' were, and still are, swiftly edited out of the discussion. The term 'gender-based violence' was adopted by the feminist movement because of the recognition that the many kinds of violence experienced by women are not only based on the biological difference between men and women, but on their socially constructed gender roles. *However*, in the context of refugee policy, this language has been co-opted, either deliberately or by default, and is often expressed merely as GBV, an easy acronym, a shorthand for all of the horror this phrase entails. It is sprinkled throughout refugee policy documents, which call for the 'protection of women and girls from GBV'. It is a form of silencing, of neutralisation, an avoidance of acknowledging the extent and the brutality of the rape and sexual violence experienced by refugee women. It denies that rape is a crime, that rape happens in all aspects of a woman's life, and that refugee women and girls are more vulnerable to rape, sexual violence and sexual torture than any other group of women in the world. *But let us be clear.* This [mis]use of language is applied universally—to the detriment of all women, everywhere. It illustrates the power of discourse to serve the powerful. The experience of refugee women has to be seen in the context of the universal experience of women and the fight against gender inequality. Because of this, we have chosen to use the terms *rape, sexual and gender-based violence,* throughout this book, without silencing the impact with acronyms. We hope that one day this will not be necessary.

Conclusion

In looking at how to scale these barriers, we need to accept that *we* cannot stop rape, sexual and gender-based violence, against all refugee women, nor insist on them being allowed to participate in the decision-making process. We wish we could! Instead, this has to be seen in the context of the global fight for gender equality, women's participation and an end to rape, sexual and gender-based violence, for all. What we can do is work to ensure that all refugee women are included in that struggle, not abandoned in the 'too hard' basket. We can challenge the structures and work with refugee women so that they can join with other women as partners. In our quest to achieve this, we explored the theories of intersectionality, identity theory, whiteness, race and social privilege (class) theory. We used these to analyse how discrimination impacts on the ability of refugee women to overcome the barriers and join all women on a level playing field[2] in the fight for gender equality and freedom from sexual abuse. These are all contentious issues and theories, and there isn't room in this book to explore all of the debates. We present how we understood and applied the theories, and how they helped us progress along the road. Fasten your seat belts for the next stage of the journey, which was more like a roller coaster ride!

References

Badiora, A., & I. (2017). Civil unrest, insurgences and the challenges of community displacement in Sub-Saharan Africa: A Silhouette of selected states in Nigeria. *International Journal of Criminal Justice Sciences, 12*(2), 302–318.

Bartolomei, L. (2015). Surviving the city: Refugee from Burma in New Delhi. In K. Koizumi & G. Hoffstaedter (Eds.), *Urban refugees: Challenges in protection, services and policy* (pp. 139–163). Routledge.

Olivius, E. (2014). (Un)Governable subjects: The limits of refugee participation in the promotion of gender equality in humanitarian aid. *Journal of Refugee Studies, 27*(1), 42–61.

Omotoso, S. A., & Fajimbola J. O. (2017). Armed conflict, terrorism and feminized poverty in developing countries: The Nigerian example. *Ubuntu: Journal of Conflict Transformation, 6*(1), 83–99.

[2] Level Playing field. A sporting metaphor in which a situation where everyone has an equal and fair chance of succeeding (theidioms.com).

Parcesepe, A., Stark, L., Roberts, L., & Boothby, N. (2016). Measuring physical violence and rape against Somali women using the neighbourhood method. *Violence against Women, 22*(7), 798–816.

United Nations Development Programme. (2016, September 14). *Under-pressure communities hosting refugees also need world's support more than ever*. UNDP. Retrieved November 10, 2021, from www.undp.org/content/undp/en/home/presscenter/pressreleases/2016/09/14/under-pressure-communities-hosting-refugees-also-need-world-s-support-more-than-ever.html

United Nations Population Fund. (2015). *For eople, planet and prosperity, UNFPA Annual Report*. UNDP. Retrieved 24 November, 2021, from www.unfpa.org/sites/default/files/pub-pdf/UNFPA_2015_Annual_Report.pdf

CHAPTER 6

Beijing + 5 the Never-Ending Story of Rape and Sexual Torture—1996–2000

Narrated by Linda Bartolomei

> *They raped us again and again. The rapes were public; they raped us in front of our children. The rapists were young rascals; try to imagine a mother raped by young boys.* (The first woman ever to testify to her experiences of sexual violence in wartime, ICTR, October 1997, in Coomaraswamy, 1999)

The next destination in the story was the Beijing Plus Five (B + 5) Review, with several side trips on the way. This special session of the UN General Assembly (UNGA) was held in New York in June 2000 to assess progress in the implementation of commitments made in the Beijing Platform for Action (BPFA). It was at this stage that I joined Eileen in this research and advocacy journey. Our aims were to monitor the impact of the promises made to improve the protection of refugee women and girls and to deepen our understanding of the extent and nature of rape and sexual torture of women in conflict and displacement.

Warning this is a very difficult chapter to read.

E. Pittaway and L. A. Bartolomei,
Only Rape! Human Rights and Gender Equality for Refugee Women,
Sustainable Development Goals Series,
https://doi.org/10.1007/978-981-19-0916-0_6

The Accidental Activist

Before continuing with the story and discussion of the achievements in this cycle of the process, I would like to take a moment to reflect on how I came to be part of this journey. I had completed a social work degree in Melbourne in the early 1980s and during that period developed a keen interest in the experiences of refugees and migrants settling in Australia. This was triggered by my knowledge of the challenges faced by my father, who had emigrated from Italy in 1959. However, during the late 1980s and 1990s, in part motivated by my frustration at what I perceived as the narrow, disempowering and apolitical frame of reference which informed much social work practice, I pursued opportunities in other fields. After travelling in Asia during the mid-1990s, I returned to Australia with a growing interest in the field of international social development. Keen to update my social work skills and qualifications in order to position myself for work in this area, I commenced a Master of Social Development at UNSW in 1997. This decision changed my life, leading to an academic and advocacy career that has now spanned almost 25 years.

It was a wonderful Master course. We were a small, tightly knit student cohort of 12, taught by a diverse team of academic and development practitioners. Our small student group agreed that Eileen was a wonderful and most inspiring teacher. We were particularly impressed by the work she had done during her years with ANCORW and her ongoing advocacy in the UN system. We loved her stories of the many hard-won achievements, but also serendipitous moments she had experienced in her advocacy work. Her frankness about how unsure and overwhelmed she had felt when she started made us feel that this was a pathway to action within our reach. Eileen was an inspiring role model, so, of course, when she accepted a full-time academic appointment at UNSW in February 1998, I jumped at the chance to have her supervise my Master thesis (and subsequently my PhD). I was invited to undertake a research project with ANCORW, to evaluate the impact of their employment programs for refugee women. This subsequently led to an invitation from the then ANCORW CEO, Velvy Holden, for me to join the ANCORW board. This I did in late 1998 and so it all began ...

Some Key Ingredients for Successful Collaboration

That's the sacred version of how and why Eileen and I started working together and, while it goes without saying that she was an inspiring mentor and that I was very much an accidental advocate and on a major learning curve, the simple truth was that we were kindred spirits. The longer we worked together, the clearer it became that our world views were strongly aligned—of course with the occasional healthy debate! and this enabled us to navigate the complexities of our changing relationship over time as I moved variously from student intern, ANCORW board member, research assistant and finally to an academic colleague and Deputy Director of the Centre for Refugee Research. Our collaboration has lasted the distance because we are both keenly aware that surviving the emotional toll of this kind of work involves regular and healthy doses of fun, never taking ourselves too seriously (not for too long anyway), celebrating the absurd and irreverent, and washing everything down with a gin and tonic. Most importantly like all good advocates, we have become adept at celebrating the small, as well as the big, wins along the way. The latter became known as our dancing moments, when we would quite literally jump up from our computers and dance around the room!

However, don't think for a minute that this story is some kind of girls' own adventure through some of the world's worst refugee camps. Nothing could be further from the truth. It was a journey so deeply traumatising that, at different times, it left us completely shattered and so overwhelmed that we struggled to function. Yet it was a journey shared, and one in which daily debriefs—over beer and hot chips in Thailand, warm gin and vermouth in Kakuma refugee camp, and illicit beers sought out in secret bars in Bangladesh—served to push us forward. As we built closer connections and strategic alliances with many grassroots refugee women's groups across the Asia region, particularly in Thailand, it was clear that we were not alone in our debrief strategies! However, the prime motivator that kept us moving forward was the gains along the way and, most importantly, the tireless determination and resilience of the refugee women with whom we worked.

Back to the Story and Progress Made Since the Beijing Conference

Following the Beijing conference (1995), significant advances had been made in the law and policy framework, in recognition of the gendered risks of conflict and the need for gender-sensitive justice. Most significant was the adoption of the Statute of the International Criminal Court (ICC) at a meeting of UN member states in Rome in October 1998.[1]

Rape Recognised as a War Crime in International Law!!

> **Rape and sexual violence were recognised as war crimes, crimes against humanity and grave breaches of human rights.** The ICC Statute listed a broad range of gender-based crimes including rape, sexual slavery, enforced prostitution, forced pregnancy, enforced sterilisation and other forms of sexual violence of 'comparable gravity' (Bedont & Hall-Martinez, 1999).

This momentous achievement was due in a large part to the tireless efforts of the Women's Caucus for the International Criminal Court, later becoming the Women's Caucus for Gender Justice, a network of women's human rights activists from around the globe lobbying for gender justice. Drawing momentum from the achievements at Beijing and the landmark prosecutions in 1998 in the International Criminal Tribunals for the former Yugoslavia and Rwanda (ICTY and ICTR), in which rape was finally prosecuted as a crime in its own right, the Gender Caucus had advocated to ensure that rape and other gender crimes in war were fully addressed in the draft statute of the International Criminal Court (Copelon, 1994).

The ICC Statute addressed many of the gaps in the treatment of sexual violence in international law. The then UN Secretary General Kofi Annan acknowledged the key contribution that the women's groups, who had advocated at the Beijing conference, had made in having the ICC established. He highlighted the importance of securing international policy commitments for action to strengthen advocacy for the critical changes

[1] The Court was permanently established in The Hague in early 2002 following the ratification of its statute by the requisite number of UN member states.

required in international law and practice. However, as we would discover, it would be many years before this important progress in international law would start to contribute to the changes needed in domestic law, policy, practice and public discourse. In the meantime, the systematic rape and sexual abuse of women and girls continued in one conflict situation after another, still failing to draw sustained international attention or condemnation. Among others, in conflicts in Indonesia and East Timor, Azerbaijan, Myanmar, the Democratic Republic of the Congo (DRC) and Sierra Leone. Time and time again, media reports of these conflicts discussed violence and killing with no mention of rape, sexual and gender-based violence. It was clear that sustained advocacy would be required at the international level if the progress made in Beijing in recognising the extent of rape of women and girls in conflict was to translate into real change.

The Rape and Sexual Torture Continue in Conflict After Conflict

> *Dozens of unknown people broke into this victim's house and plundered it. They stripped off her dress, raped her and mutilated her vagina with barbed wire. This victim died in hospital* ('L' – 11 years old, TRuK, 1998, p. 6).

In September 1998, Eileen had attended an Asian Women's Human Rights Council Meeting held in Manila (AWHRC). At that meeting, Indonesian Council member, Ita Nadia, had shared horrendous accounts of rape and abuse of women during the so-called May riots, perpetrated by the Indonesian military a few months earlier. During these brutal attacks by militia-like gangs, over 1000 people were killed, at least 152 Indonesian Chinese women were raped, and hundreds of Chinese-run businesses and homes were razed to the ground. Most rapes reported were gang rapes involving large groups of young men.

> **Rape as Torture**
> *Patterns of sexual mutilation of women and girls' breasts and vaginas using barbed wire, glass and teeth were also documented. In addition to raping the women, the rapists severed women's nipples with wire cutters, shot rubber*

> *bullets into their vaginas and covered their heads in plastic bags before attempting to drown them.* (Pittaway & Bartolomei, 2005)
>
> *After the two girls got away from the savage rapists, I came to them and embraced them. They begged me to help them find a secure way to get home. Since I live in that district, I know well the short-cut to the main road. At the Cengkareng crossroad, I saw some nude female corpses, their faces covered with newspapers. They must have been raped, for I could see the dried blood on their vaginas on which the flies swarmed.* (TRuK, 1998, p. 1)

The rapes reported detailed shocking levels of brutality which could only be described as sexual torture. Two of the young women had sustained such serious injuries that, without urgent medical aid, it was clear they would die. Support was sought from neighbouring countries with well-developed emergency evacuation procedures and advanced medical systems, but for political reasons, none agreed to help. The two young women subsequently did die from their injuries. Human rights groups across Indonesia had placed responsibility for these attacks firmly at the feet of the Indonesian military, but this was denied by the Indonesian Government (JFFT, 1998). However, because women's rights activists had spoken publicly about the rapes, both they and their children were threatened by the Indonesian military (Human Rights Watch, 1998). This prompted them to call for assistance and support from human rights organisations outside Indonesia. Ita asked members of AWHRC to support in any way possible.

Eileen returned to Sydney from the AWHRC meeting, completely shattered, but determined to assist Ita in her struggle for justice for the Chinese women. At the next meeting of the ANCORW board, the first I attended as a newly minted board member, it was the top item on the agenda. The board listened in horror as Eileen recounted what she had learned from Ita. I could not recall ever having heard stories of such brutality and I was distressed and angered by the clear lack of accountability by both the Indonesian Government and the international community. I felt morally obliged to respond to Ita's call for help. Several others on the board felt the same. A sub-committee, comprising Eileen, Cate Breen, Yumi Lee and I, was soon frantically planning a research and advocacy strategy. The first stage of this strategy focused on assessing the extent of the media coverage of the rape and sexual violence which had occurred.

Rape Reporting: Silenced by the Media

Two social work students completing internships with ANCORW in early 1999 assessed the extent of media coverage of the rapes that had occurred in Indonesia. They searched all the major Australian newspapers over a six-month period but found only a single reference (Greenlees, 1998). The issue was never front-page news and received no sustained coverage. Similarly, our attempts to interest the Australian media all failed. Journalists were not interested in reports of the rape and abuse of women unless a woman prepared to share her first-person account could be produced (Journalist 1 Pers. Comm February 1999). Another journalist writing on international affairs for one of the major newspapers indicated his lack of interest in the story unless it highlighted the systemic failures of the UN. The widespread rape and abuse of women were, in itself, not interesting. His final remark only served to highlight the ongoing battle against institutionalised misogyny and the ways in which crimes of sexual violence continued to be reduced to at, best, bad behaviour:

> 'So, is rape really a war crime now? Remind me to never join the army!' (Journalist 2 Pers. Comm March 1999).

We were deeply disturbed by the attitudes underpinning the lack of media interest. It was against these ideological and attitudinal barriers that the survivors of rape and sexual torture, and the grassroots NGOs who supported them, sought to be heard. It was clear that the inclusion of rape as a war crime in the Statute of the ICC, while momentous in our minds, was still having limited impact on changing media, public discourse and social attitudes.

Building the Strategic Advocacy Campaign: Expanding Our Support Base

Having failed to engage the interest of the Australian media in the struggle, we reviewed our options and strategised other means that could be used to raise local and international awareness of the issue. We focused our efforts on two complementary advocacy strategies. The first was to build support among the women's rights NGOs across the Asia–Pacific region to take this as a major issue to the forthcoming B + 5 review, to be held in New York in June 2000.

The Importance of B + 5, 2000: A High-Level Meeting and Special Sitting of the UN General Assembly
Held five years after the Beijing conference, this session evaluated progress and identified pervasive gaps in implementing the commitments made in Beijing. It was a specially convened sitting of the UN General Assembly, led by then UN Secretary General Kofi Annan. This gave the meeting and the Political Declaration and Outcomes Document, agreed by consensus, a very high international standard-setting status. The special session attracted ministerial level participation, with all states expected to report on their progress in implementing the Beijing Platform for Action (BPFA).

In the lead up to the B + 5 review, Asia–Pacific Women's Watch (APWW), of which Eileen was co-chair, brought together women's rights activists from across the Asia–Pacific region to plan their input into the meeting. Preparatory meetings were held during 1998 and 1999 and Eileen's position provided important benefits, including access to funding to attend regional preparatory meetings, and the role of over-seeing the preparation of all the advocacy documentation and meeting reports. She was therefore well placed to not only represent ANCORW at all the key UN meetings, but also had significant power to ensure that our advocacy concerns were well represented in all the reports and advocacy documents.

At this point, my role was a purely voluntary one, fitted round my work as a casually employed community worker. I did not for one minute imagine that I would actually participate in the meetings in New York the following year. When this happened, it was some time before I actually felt comfortable to remove my self-described label of 'accidental activist'. The fact is I was incredibly naïve and totally inexperienced at advocacy in the UN system. Nonetheless, I was driven by the strong desire to make a difference in some small way.

Mobilising Support at the Sub-Regional Conference, Sydney, July 1999

In July 1999, activists travelled from across Australia and our sub-region to a conference we convened in Sydney, hosted by ANCORW and Asia–Pacific Women's Watch (APWW). The purpose was to develop strategies across all twelve areas of the BFPA for B + 5 meeting. ANCORW and APWW held a workshop at which refugee women, resettled to Australia from diverse conflict-affected countries, shared accounts of systematised rape, sexual abuse and violence. They highlighted the risks that displaced women and girls face in every aspect of their lives, both in armed conflict and in refugee sites. They confirmed that the extent and impacts of this abuse were still rarely acknowledged. Seldom, if ever, were the devastating impacts of rape and sexual violence, including the many babies born of rape, recognised or discussed. Women named the multiplicity of perpetrators including the military, police, border guards, humanitarian workers, men in host communities and men in refugee communities. They emphasised the fact that while in many of the conflicts, rape was a planned and systematised military strategy; this was still not recognised by the international community. Rape in war was still often ignored and not taken as seriously as other war crimes.

What Is Systematised Rape?
The soldiers came to our village and took away all the men and boys. They then made the women and girls go into the barracks where they were all raped, even the young girls. They then took them outside to where all the men and boys had been shot dead and castrated, and the commander said 'Have a good look, they will never f.... you again. Now tell the others that this will happen to them next', and all the soldiers laughed. The women will never get over it. Some want to take their own lives (Testimony from a Refugee woman, 2000).

The refugee women made strong recommendations on the importance of all governments signing the Statute of the ICC and of restating the commitments they had made in Beijing to recognise and address systematic rape. This, they argued, continued to be one of the major barriers faced by refugee women in their struggle for gender equality. Their recommendations also emphasised the importance of participation, and of *all* governments recognising rape as persecution and as grounds for refugee status. They gave their full support to ANCORW to take this as a priority issue to the regional APWW meeting to be held in Bangkok in September 1999. Our advocacy objective, to build a wider support base for advocacy at the sub-regional level, had been achieved. The next step was to mobilise the same depth and breadth of support at the forthcoming Regional NGO meeting in Bangkok.

Using Film for Advocacy

The second advocacy strategy was to make a documentary film to highlight the extent and impact of rape and sexual violence in conflict, and to document the advocacy campaign in which we were involved. The power of the online world, YouTube and social media was yet to emerge, and thus, the use of film for advocacy was well received as an innovative and powerful way in which to bring the voices of refugee women to policy-makers, governments and UN officials who held significant power to bring about change.

The aim was for the film to bear witness to the atrocities that the women in Indonesia and other contemporary conflict and displacement situations had experienced, and to document their struggle for justice. It had become increasingly clear, as the workshop with refugee women held at the sub-regional conference had highlighted that the rape of women in Indonesia, the denial by the Indonesian Government and the lack of sustained interest on the part of the international media were far from isolated incidents. Instead, they formed part of a larger international pattern of 'silence' in which, despite the progress made during the previous decade, the extent of rape and sexual violence in conflict and refugee settings was still rarely acknowledged or addressed. The struggle to have the endemic rape perpetrated in conflict acknowledged, let alone prosecuted, was a core part of the struggle.

Two committed feminist community film makers, Robyn Kennedy and Anne Morphett from Critical Distance Films, eagerly agreed to assist and the process of filmmaking commenced. This proved to be the beginning of a long and difficult process of fundraising and filming which stretched over a two-year period. The result was a one-hour documentary, *Only Rape*[2] (Pittaway & Bartolomei, 2005). The film commences with the launch of the advocacy campaign in Sydney and documents the incidence of, and response to, rape in Indonesia, as well as in many other conflict settings around the world. It charts our advocacy journey through multiple regional and UN meetings over the period discussed in this chapter.

Building Our Support Base at the Regional NGO Conference, Bangkok, September 1999 Expanding the Evidence

Eileen, with ANCORW board member Juliana Nkrumah, attended the Regional Conference and convened a workshop on rape and systematised sexual torture in conflict and refugee settings. It attracted the attention of women, and some men, from across the region, all keen to share their stories. Several Indonesian women's rights activists and women from Aceh and East Timor, gave testimony to the workshop (Pittaway & Bartolomei, 2005). Participants listened in horror and disbelief, both men and women, as story after story was told. The stories clearly identified how rape is used as a form of torture and as a strategy to demoralise and destroy refugees' lives and communities.

Rape as Torture
An older male human rights activist fought to speak through his tears as he told of an event he had witnessed. '*The army forced a son to rape her his mother in front of so many people and nobody could help her. This happened in Aceh*'. He described how the villagers were forced to watch at gunpoint, and how the young man howled in distress.

The women from Indonesia confirmed that, in spite of some brief international media coverage of the rapes, and an investigation by the then UN Special Rapporteur on Violence against Women, Radhika

[2] Due to lack of funding, the film was only completed in 2005. John Reynolds and Jill Hickson, Art Resistance Films, worked with us to produce the final product using the footage provided by Robyn and Anne.

Coomaraswamy, no sustained interest had been shown by the international community. An investigation, undertaken by the Indonesian Human Rights Commission, had been dismissed by the Indonesian Government, who denied that the rapes had ever occurred. The human rights activists who had documented the rapes continued to face threats to remain silent or face attacks on themselves and their families (TRuK, 1998). Similarly, when the UN Special Rapporteur's report was tabled at the Human Rights Council in April 1999, it was rejected by the Indonesian Government (Pittaway & Bartolomei, 2005). The discussions in this workshop again highlighted the multiple political barriers in the struggle for justice for the women in Indonesia, and in other conflict settings across the region and the globe, who had been raped and sexually tortured in conflict. With the permission of the workshop participants, the workshop and the women's accounts of rape and sexual abuse were recorded on film as well as included in the workshop report.

Eileen led the drafting process for the report produced from this Regional Conference. It included a strong focus on the impact of rape and sexual torture in refugee and conflict situations. It also emphasised the importance of refugee women's leadership and participation in the development and implementation of solutions. The report identified a series of advocacy strategies to address these human rights abuses and to bring the perpetrators to justice. Building on the advocacy strategy used in Beijing, it was accompanied by a booklet entitled *Voices 2000* (APWW, 2000), the 'Little Blue Book'. In this booklet, the key issues of concern were summarised and accompanied by clear recommendations for action. Women from across the Asia–Pacific used this to lobby governments at the formal Preparatory Committee for B + 5 in New York in early 2000, at which the text of the Outcomes Document for the B + 5 meeting was being debated.

Playing Politics: Mobilising the Support of the Australian Government

As we called on women across the region to lobby their governments to make strong commitments to recognise and address rape in conflict and refugee settings, we tasked ourselves with mobilising the support of the Australian Government. In the intervening years between the Beijing Conference in 1995 and the planned B + 5 review in 2000, there had been a change of government. Where the previous Minister for Women had been a strong ally of the women's movement, the new Minister for Women was less supportive. The women's movement had learned a salutary advocacy lesson in these years. When they had sought a meeting with the newly appointed Women's Minister, they received a less than enthusiastic welcome. She made it quite clear that their failure to engage with her during her time in opposition had been a major mistake. Keen to not repeat the errors of the past, we worked hard to build strong links with both the key policy-makers in the Office for the Status of Women (OSW) and female politicians across the political spectrum.

In November 1999, with the Regional Conference report in hand and a five-minute excerpt of the film of the workshop held in Bangkok, we travelled to Canberra to meet with the cross-parliamentary women's committee. Our aim was to build strong bipartisan support for the Australian Government delegation attending B + 5 to articulate strong commitments to addressing rape and sexual violence. This proved to be a very effective strategy. The representatives from OSW, who were involved in meetings to draft the Australian input to B + 5 Outcomes Document, became some of our strongest allies and the source of very useful inside information.

Preparatory Committee for B + 5 CSW March 2000

The next key event was the meeting of the Commission of the Status of Women (CSW) in March 2000, which also served as the official Preparatory Committee for the B + 5 Review. Eileen had been appointed to the official NGO Organising Committee, CONGO, and she was funded to be in New York for a month. This was a powerful position from which to influence the agenda of the NGO advocacy community. Without funding, and staying in the cheapest hotel we could find, the filmmakers and I could only afford to join Eileen for the second half of the meeting. Shortly before we left Australia, Eileen called me to restate her earlier advice that I should make sure to sleep on the plane so that I could hit the ground running in New York. She was not wrong. I arrived a few days later, the film makers in tow, and quickly settled into a routine of early morning strategy meetings, followed by a sprint to Kinko to photocopy our advocacy documents, long days wandering the labyrinthine corridors of the UN and dashes across the road to the Church Centre to attend NGO caucus meetings. The Church Centre for the UN was an important hub for meetings of religious organisations and NGOs working for peace and human rights. I can hardly complain, as Eileen did all of that and was then called into either CONGO or APPW meetings every evening, leaving me with a few precious hours to join the filmmakers and explore New York. Breakfast and lunch were grabbed on the run, if at all, and frequently the words of Caroline Lambert, one of our advocacy partners, highly skilled in UN advocacy, rang in my ears. She had said that the most important thing to remember was to *'wear comfortable shoes and keep an emergency supply of Gummy Bears in your pocket'*!

Joining with the NGO Women's Caucus on Armed Conflict: The Never-Ending Story of Rape Continues

At the CSW meeting, we joined with a group of women's NGOs who were lobbying for a greater recognition of the impact of armed conflict on all women and for the active inclusion of women in peacebuilding processes. Led by Felicity Hill, a representative from the Women's International League for Peace and Freedom (WILPF), this group worked as the NGO Women's Caucus on Armed Conflict. They convened daily strategy meetings and jointly lobbied governments and UN officials. We heard, and met, women from around the world who were advocating on the same issues. They shared detailed accounts, often with chilling similarities, of the way in which rape had been, and continued to be, used as torture and a systematic strategy of war. In the conflict in Sierra Leone in the previous year, the rape of women had been used as part of a planned strategy by the rebel soldiers to break the morale of the population of the city of Freetown and to diminish the possibility of an opposition.

'Go with the rape first'!
Isha Dyfan, from WILPF, graphically described the scale of rape of women in Sierra Leone that had occurred on the 6th of January 1999 as the rebel soldiers had swept into Freetown. The soldiers were directed to 'go with the rape first'. She explained that they had used rape as the strategy by which they would enter and dominate the city. She described women held in sexual slavery in rape camps where they were gang-raped repeatedly and kept permanently naked to prevent their escape (Pittaway & Bartolomei, 2005).

Jessica Babihuga Nkuuhe from Women's International Cross-Cultural Exchange, Uganda, shared her analysis. She spoke of the impact of shame and stigma on the women who survived and, in particular, on those who gave birth to babies of rape, noting that, *'the silence around these issues is unimaginable'*. Elmira Suleymanova from Azerbaijan gave accounts of the number of women who had suicided to avoid being raped. She reported that, in one day alone, 70 young women had taken their own lives. She described her work to support women who had been raped in refugee camps and the despair she felt when they were rejected by their families and communities (Pittaway & Bartolomei, 2005). International NGO colleagues had also shared credible reports of a manual from the Burmese Military which detailed how to use rape in conflict. It was clear that in conflicts in every region of the world, the systematised rape and sexual torture of women and girls continued unabated. As Eileen said at the time, it often felt utterly overwhelming, as we would hear about the rape and sexual torture of women in one conflict and then another and another and another...

Towards Security Council Resolution 1325 and a 'Me Too' Moment

One of the main advocacy aims of the Caucus on Armed Conflict, chaired by WILPF, had been to have women's concerns in armed conflict addressed within the formal agenda of the UN Security Council. Following intense lobbying, the first step in this process was taken on International Women's Day in March 2000.[3] The UN Ambassador for Bangladesh, the then Chair of the UN Security Council, had announced his government's commitment to convening a special session of the Security Council to consider the impact of conflict on women, and women's role in peacebuilding. This announcement was certainly one of our dancing moments! I had not yet arrived in New York to join the CSW meeting, but was thrilled to receive a very excited call from Eileen to share the news. The filmmakers were keen to include this landmark event in the film. When they arrived in New York, they arranged a meeting for us with a senior government delegate on the Security Council. He

[3] Security Council Resolution 1325 | Peace Women (http://www.peacewomen.org/SCR-1325).

enthusiastically discussed the urgency of convening a special session given both the impact of conflict on women and girls and their important role in peacebuilding. His enthusiasm for the issue was only exceeded by the frequent and very obvious glances at my cleavage. When, as the meeting ended, he discovered that I would be staying in New York after Eileen and the filmmakers had left, to attend the preparatory meeting for the ICC, he proceeded to ask me out to dinner! The filmmakers were flabbergasted by this exchange, and we later wryly joked that an alternative film title might have been *Linda's Cleavage Saves the World*. Deep down, we were far from flippant about this exchange. It served to highlight, as had Eileen's similar encounter with an Australian Government bureaucrat, that gender inequality was deeply entrenched at all levels and had to be challenged if we were to make any progress. However, it was not just powerful men with whom we would struggle during the B + 5 process. Increasingly, the threats would come from a toxic mix of patriarchal, right-wing and extremist religious groups, involving both women and men.

Ugly Ideological Clashes

During the CSW meeting, we had been utterly astounded by the enormous number of mainly white, American right-wing Christians who had participated in the meeting. Hundreds of earnest young white men in suits and women in demure dresses would descend on the UN building each day. They were joined by priests, nuns and a core of well-dressed and very articulate middle-class women. It quickly became clear that their mission was to ensure that the Outcomes Document included no references to sexual and reproductive rights. In plain speak, that meant no support for abortion or contraception and no rights for lesbian or transgender women. Their tactics were familiar, as any woman running the gauntlet to an abortion clinic well knows, and very nasty. They walked the long, and often dim, corridors of the UN in large groups, seeking out women in traditional dress from the global south. Upon approach, if the woman made it clear that she did not support an anti-abortion agenda, they would surround her, kneeling and loudly praying for her soul. They particularly targeted meetings of the Lesbian Caucus. This included positioning tall men in cassocks, wearing huge crucifixes, at the door to the meeting in order to intimidate the women who wished to attend. They had also lined the room with young white men who stood, rosary beads in hand, at the rear of the room. It was all like a scene from a bad remake of the film *The Exorcist*, but then it became even more surreal. At one of the meetings, one of the speakers was a woman from Canada representing Catholics for a Free Choice. As she took to the floor to speak, the young rosary bead-wielding men began to quietly chant 'burn her, burn her, burn her'! Their level of coordination, steely determination and absolute belief in the righteousness of their agenda was utterly terrifying.

'Babies of rape are a gift from God'
Another deeply disturbing event occurred during an encounter with a woman from a Central American country who was providing legal counsel to the Holy See.[4] We had approached her, in the misguided belief that the Holy See would be supportive of our advocacy to have rape prosecuted as a systematised strategy of war. Her immediate response was that the reports of rape of women in war had been vastly exaggerated. When we challenged this, pointing to the recent conflicts, including Sierra Leone, where the huge numbers of women had become pregnant from rape, she then changed focus and asked us our position on access to abortion. We replied, of course, that our position was one of free choice and that access should be particularly made available to women and girls who had been raped. Her colleague jumped in with a clearly well-rehearsed retort. 'What we had to understand' she said, *'was that the baby was a gift from God to make up for the rape'*. We later learn from the Gender Caucus for the ICC that this had been a major flashpoint during the negotiations for the Rome Statute of the ICC. The Holy See, supported by a number of anti-choice organisations, had actively obstructed the definition of forced pregnancy, fearing that this would provide a justification for the widespread legalisation of abortion.

Struggling to Maintain Commitments to Advance Refugee women's Rights

Resisting the very destructive impact of these right-wing Christian groups consumed an enormous amount of energy for the women's movement and key NGO organisers. Worst of all, their efforts, fed by emerging global conservativism, and a decreasing willingness on the part of

[4] The Vatican.

major states to commit to strong human rights-based language, seriously impacted progress on women's rights at the CSW meeting (Bunch, 2000). We, and others in the women's movement, were becoming increasingly concerned that not only might forward progress be impossible, but that advances made in the human rights language in Beijing might also be under threat. In the last days of the meeting, despite much of this having been agreed at the Beijing conference, the language relating to the rape of women in conflict situations and to the protection of refugee women remained in square brackets. This meant that consensus on its inclusion in the report had not been reached among the UN member states. As the CSW meeting concluded, it was clear that the B + 5 June meeting was going to be similarly long and contentious. Given the very limited progress that had been made drafting the Outcomes Document, an additional week was added to the two week-long meeting in June. This surprised everyone as it had been expected, as in previous UN meetings, that the draft Outcomes Document would have been fully negotiated at the CSW meeting. The Special Session of the General Assembly for B + 5 in June should have been a largely symbolic process in which the Political Declaration and Outcomes Document were formally approved.

Neither Eileen nor I was prepared for this. I had used the little money I had, and borrowed extra, to self-fund this trip to New York. I had not anticipated attending the June B + 5 meeting. Eileen was funded through her role on CONGO to attend, but given her university teaching commitments could not arrive in New York in time to participate in what was clearly going to be an additional crucial week of advocacy. However, *we had to be there!!* We returned to Australia determined to find a way. The 'bank of family' covered my return airfare to New York.

Ongoing Challenges for the Women at Risk Program

Meanwhile, back in Australia in early 2000, ANCORW had identified that the WaR Program was again struggling to fulfil its potential (Hercus et al., 2000). In 1999 and 2000, the Program had fallen significantly short of its target to allocate 10.5% of Australia's resettlement places to WaR. In 1998/1999, 72 places had remained unfilled, and in the following year, this had risen to 135 places (DIMIA, 2000). The Australian Department of Immigration (DIMIA) and UNHCR staff had also expressed concerns about whether the program was targeting those most at risk. Two key problems had been identified. The first was confusion over the definition and interpretation of the term 'women at risk' by staff who determined the eligibility of women for the program, and second were assertions that some refugee women were manufacturing claims of rape and sexual violence in order to be accepted into the program (Hercus et al., 2000). These findings directly informed our advocacy strategy at B + 5 and further research in the next research and advocacy cycle.

Progress at B + 5 and the Importance of Insider Status

By the B + 5 June meeting, there was a detailed draft of what would become the final Outcomes Document, still with many square brackets. A key part of CONGO's role was to provide daily morning and afternoon updates to the NGO sector on the progress of negotiations between member states towards the final text for the Outcomes Document. This gave us the opportunity to share our advocacy positions and seek strategic support from NGOs from around the world to lobby their governments to support them. By the June meeting, Eileen was appointed to lead the daily afternoon briefings and nominated me, along with several NGO colleagues who had been closely following the negotiations, to provide a daily update on the progress of the Document. This further increased the reach and impact of our advocacy, for, while we provided reports on all the key women's human rights issues under discussion, we never missed a chance to seek further advocacy support for text addressing systematic sexual and gender-related violence. Day after day, we shared a forensic line-by-line analysis of the relevant text, highlighting the key sections on which governments were yet to agree and urging the women in the meeting to keep up the advocacy pressure on their governments. We celebrated any small progress forward and were particularly pleased by the number of paragraphs that referenced both the particular barriers that refugee women faced (Paras. 5, 15, 16, 17) and the importance of their full and equal participation (Paras. 99g & i).

Before the Days of Digital and Online Advocacy

As these were still early days for mobile phones and ready access to WiFi, we remained dependent on hard copy reports and documents to support our advocacy. We frequently rushed to the nearby photocopy shop to make multiple copies of documents and advocacy fliers or we sneaked into the UN library to use their machines. The need to check a policy or legal precedent to reference our lobbying would send one of our wonderful social work interns running to the library to scribble the reference on a piece of paper so that we could then pass it discreetly to sympathetic government representatives engaged in the drafting process.

The last few days of the meetings seemed interminable. Day after day, night after night, the meetings dragged on, as governments were unable to come to an agreement on the wording or many key paragraphs in the document. Much of the consensus and shared vision, which had enabled such a strong document to be produced at the Beijing conference, had disappeared. Governments had become increasingly cautious about agreeing to anything that might imply increased accountability for human rights abuses or increased funding obligations. Reproductive and sexual rights remained a key flash point. Our advocacy became increasingly strategic as we directly targeted sympathetic members of government delegations, providing alternative language directly linked to the paragraphs in the draft Outcomes Document.

Refining Our Advocacy Focus to Persuade States to Recommit to Beijing Language

We narrowed our focus to lobby on two issues, against significant resistance from many states who were seeking to 'step back' from some of the human rights commitments made in Beijing. The first was for governments to restate their recognition of rape as a widespread and systematised strategy of war, and to commit to signing the ICC Statute. The second was for explicit recognition that rape, sexual and gender-based violence lead to the stigmatisation and social exclusion of women and is perpetrated in war with this express intention. It therefore should be accepted as grounds for refugee status and for resettlement under the Women at Risk Program. While a small number of states had started to accept that rape and sexual violence constituted grounds for refugee status, this was far from universal.

It also continued to be a major barrier to the implementation of the WaR Program (ANCORW, 2000).

After much lobbying, a reference to the systematised nature of rape was included as one of the obstacles to the protection of women in armed conflict in the draft Outcomes Document (UN, 2000, para. 19). We had been working late in our hotel room when we received a call from one of our allies in the Australian Government informing us of this. By happy coincidence, the filmmakers were also in the room, so the telephone exchange was all caught on film, providing good insight into the politics of the lobbying process. However, advocacy relating to the recognition of rape as grounds for refugee status proved far more challenging. In spite of language in the BPFA (UN, 1995, paras. 147h & i), which had encouraged governments to consider recognising women who had experienced sexual violence or other forms of gender-related persecution in conflict as refugees, and the application of this principle in some asylum decisions, strong resistance to committing to this remained (Pittaway & Bartolomei, 2001). We were advised by one of our friends in the Australian Government that most developed country governments, including Australia, were terrified of making these commitments, fearing that it would open the flood gates to many more asylum applications. In fact, the resistance was so great that the inclusion of any language which urged governments to take further steps in this regard seemed increasingly unlikely.

They Were Wrong! an Advocacy Win for Refugee women's Rights

On the penultimate day of the B + 5 meeting in June, the language relating to gender-related persecution and asylum was still in square brackets. Desperate to complete the negotiations, governments and advocates had worked until 2am the night before. As negotiations had become so contentious, particularly those relating to sexual and reproductive rights, governments broke into a number of informal closed sessions to complete their negotiations. NGOs were not allowed into the closed sessions, so we had to rely on supportive government delegates. We sat in the corridors waiting for news, only to be disappointed when at nearly

3am, our friend from the Australian Government emerged to say that there was still no agreement regarding the wording on gender-related asylum. We dragged ourselves back to our hotel, weary and dejected, slept badly and were back at the UN for the 8am NGO briefing the following morning.

Throughout the day, in the corridors, the coffee shop, even the bathrooms, we lobbied every NGO and government delegate we could find, but were not hopeful of success. In an attempt to cheer ourselves up, we sat with the filmmakers in the UN cafeteria for a glass of wine as we watched the sun sink over the Hudson River. We told them to pack their equipment and head out to enjoy an evening in New York as we did not anticipate there would be anything worth filming that night. But just in case, the producer insisted we keep a hand-held video camera with us and quickly showed our social work intern, Maggie, how to use it. What a fortuitous decision! Against all odds, in the early hours of the morning, as we sleepily listened to the final draft of the document being read out, we were stunned to hear that those few words that had been the focus of days of advocacy had finally been included in the document!!

The Inclusion of Our Language Against All Odds!
The B + 5 Outcomes Document states the that governments should: 'Mainstream a gender perspective into national immigration and asylum policies, regulations and practices, as appropriate, in order to promote and protect the rights of all women, including the consideration of steps to recognise gender related persecution and violence when assessing grounds for granting refugee status and asylum' (UN, 2000, para. 86i).

This paragraph provided an important acknowledgement that the lack of recognition of rape and gender-related violence as torture remained a barrier to the protection of refugee women. It was also a symbolic hook on which to hang future advocacy for the universal recognition of gender-related persecution as grounds for protection within the terms of the UN 1951 Refugee Convention and, importantly, for resettlement under the WaR Program. With this in mind, our weary cheers resonated as we danced down the long, dark, almost empty corridors of the UN at 4am on Saturday 10 June 2000. As Maggie captured our excitement on film, the sun slowly rose over the Hudson River. Our dejection at the setting of the sun had changed to euphoria as it rose on a brand new day.

From Survivors to Contributors: Recognising Refugee Women's Capacities

Following sustained advocacy from the women's movement led by WILPF, in October 2000 the Security Council convened a special session on women and peacebuilding. The outcome, UNSC Resolution 1325 on Women and Peacebuilding, was adopted on 20 October 2000 (UN, 2000). This ground-breaking resolution significantly advanced the struggle for the human rights of refugee and conflict-affected women around the world. It was the first time the Security Council had addressed both the disproportionate and unique impact of armed conflict on women, and their contribution to peacebuilding, which, prior to this, had been undervalued and underutilised. This legally binding resolution clearly acknowledges the gendered impact of war and placed a number of obligations on governments and other parties to conflicts. These include the obligation to pay attention to protecting women and girls from rape and sexual abuse, a particular focus on the needs of refugee women and girls and, most significantly, the pivotal role that women can, and should, play in leadership and decision-making. It stresses the urgency of involving them as equal participants in all peacebuilding activities.

UNSC Resolution 1325 on Women and Peacebuilding
'Calls on all parties to armed conflict to take special measures to protect women and girls from gender-based violence, particularly rape and other forms of sexual abuse, and all other forms of violence in situations of armed conflict (UNSC 1325 para. 10).

Calls upon all parties to armed conflict to respect the civilian and humanitarian character of refugee camps and settlements, and to take into account the particular needs of women and girls, including in their design'... (UNSC 1325 para. 12).

'Urges Member States to ensure increased representation of women at all decision-making levels in national, regional and international institutions and mechanisms for the prevention, management, and resolution of conflict' (UNSC 1325 para. 1).

As Security Council resolutions are binding on all UN member states, this was a significant milestone in the ongoing struggle for the human rights of refugee women and girls. It had been built directly on commitments, made in both the Beijing Platform for Action (1995) and the

B + 5 Outcomes Document, to refugee and conflict-affected women. After years of struggling just for the extent of rape and sexual violence in conflict to be acknowledged, the obligation set out in this Resolution, to support women's participation as a central part of prevention and protection, was a major step forward. Finally, refugee and conflict-affected women were being viewed, not as helpless victims requiring protection and care, but as people with skills, capacities and knowledge to contribute to humanitarian responses and peacebuilding. After years of rhetoric on the importance of refugee participation, this was indeed progress and a significant milestone in the ongoing struggle for refugee women's rights.

Moving Forward—Lessons Learned in This Research and Advocacy Cycle

In this cycle, we strengthened our advocacy by drawing on the work of key feminist scholars in the field of development and refugee studies. Our academic learning was informed as much by the discursive environment of the women's human rights movement across the Asia–Pacific as it was by theoretical readings. Through our involvement in women's conferences in the lead up to and at B + 5, we met, and learned from, many incredible feminist scholars and thinkers[i] from across the global south. Their analyses and critiques challenged traditional Western feminist approaches and understandings of women's oppression that had long ignored issues of power, race, politics and colonisation. Within a southern or post-colonial feminist critique, these were recognised as central concerns in the emancipation of women.

They argued that much feminist scholarship had privileged women in the global north as the only legitimate subjects of the feminist struggle and had marginalised the voices of non-white women. In the global feminist movement, women from the global south were heard only as 'fragmented, inarticulate voices in (and from) the dark' (Mohanty, 1988, p. 193), and it was argued that feminism did not typically consider the impact of race, class, religion or ideology on individual women's lives (Mohanty 1991, p. 31). Women activists from the global south called for a critical feminist scholarship which would recognise and privilege the voices and capacities of women of the south. They argued that the homogenisation of all women in the south as oppressed, obscured their multiple identities and capacities, and prevented recognition of the places where they were able to wield power as mothers, sisters, wives and community providers. This is when we first started to think more deeply about the additional barriers that refugee women and girls face in the struggle for gender equality.

Refugee Women and Agency: A Gendered Analysis

This characterisation of women from the global south as vulnerable and powerless, coupled with the identified challenges in involving them in decision-making, was increasingly being discussed in the post-colonial feminist literature. Key scholars had critiqued the way in which refugee and so-called Third World women were homogenised and denied a voice as a result of colonial development approaches, which tended to be paternalistic and assumed that 'expert' outsiders (usually white) understood the solutions to the women's problems better than they did themselves (Mohanty, 1991; Ong, 1988; Spivak, 1988). Their analysis highlighted that these approaches both increased women's risks of human rights abuses and contributed to their exclusion from decision-making processes. This post-colonial feminist critique of development approaches had also begun in the late 1990s to influence the field of refugee studies. From this time, a number of writers had highlighted the way in which refugee women had so often been characterised within traditional aid and development discourses as hopeless, vulnerable and desperate (Salinas, 1988; Schmuki, 1998). As Doreen Indra (1999) argued, these stereotypical images of the desperate and vulnerable refugee women of Africa meant that they were never considered as anything other than desperate aid-dependent victims. In this way, traditional models of aid delivery, rather than protecting the rights of refugee women, had instead reinforced their oppression and exclusion. Many humanitarian actors continued to take a 'vulnerability'-based approach to programs for refugee women, which portrayed them as 'powerless, helpless needy people' (Coomaraswamy, 1999). The lack of women leaders in refugee settings and the inclinations of the local male UNHCR and NGO staff to reinforce patriarchal approaches only further increased women's marginalisation. For example, all over the world a key part of women's role is accessing and providing food for her family, yet in many refugee settings food distribution was controlled by police, military and male leaders. This led to another site of sexual violence as some men demanded that women exchange sex for their food rations. The participation of women except as beneficiaries was still resisted by many UN and NGO development practitioners (Indra, 1999; Parpart, 1995).

These women challenged the way 'knowledge' about refugee women was produced and filtered through organisations and institutions, thereby denying refugee women voice and agency. They outlined the urgent need for grounded, reciprocal and locally based research and training initiatives to challenge the limitations of traditional development approaches and to genuinely support the implementation of gender-sensitive and participatory approaches in work with refugees. Refugee women themselves had also begun to challenge their characterisation within traditional aid and development discourses as hopeless, vulnerable, weak and entirely dependent on aid for survival. In her research with Tamil Sri Lankan women, Schrijvers (1999) highlighted their strengths and capacities against a backdrop of international aid provision which highlighted only refugee women's poverty, fragility and vulnerability. She argued against their stigmatisation as vulnerable, pointing out that they are at once both vulnerable and strong (Schrijvers, 1999).

Drawing on the Southern Feminist Critique in Our Advocacy

It was this game-changing analysis which had underpinned the advocacy at B + 5, for UNSC 1325 and to secure strong commitments to supporting women's leadership and participation. Despite this need being first recognised at the third World Conference on Women held in Nairobi over 15 years earlier, it had been a protracted policy failure. Since the Nairobi Conference, there had been repeated calls for the increased participation of refugee women in decision-making and camp management (Forbes-Martin 1992; Kelley, 1989), and it was emphasised in numerous UNHCR policy documents which recognised women's participation as a critical protective factor in reducing the risks of rape and sexual violence and in aiding recovery (Chester, 1992). Yet in 1992, Susan Forbes-Martin had noted that 'refugee participation probably has the worst ratio of rhetoric to reality of any concept in the refugee field' (Clarke 1987 in Forbes-Martin, 1992). It was highlighted again in 1995 at the Beijing Conference (UN, 1995, para. 147a) and again in 2000 in the B + 5 Outcomes Document (UN, 2000, para. 99g). It was its inclusion in the binding Security Council Resolution 1325, on the back of feminist critique from the global south, that had finally signalled real progress.

Conclusion

We were very proud of what we had achieved in our advocacy at the B + 5 Conference. Once again, we had felt extremely privileged to have had the opportunity to work with so many skilled and internationally renowned women's rights advocates, and to have been able to play a small part in contributing to the positive progress made in incorporating gender into international law and policy. However, as our euphoria started to dissipate, we paused to consider what real impact these victories might have on the lives of the women who had been raped in Indonesia or for the Sierra Leonean women now displaced in refugee camps in Guinea and Liberia. How long would it take for these commitments to be implemented in policy and practice? As we analysed the outcomes of B + 5, it was clear that the failure of governments to agree to additional 'concrete benchmarks and time-bound targets' for the implementation of BPFA commitments was a major concern (Girard, 2000, p. 1). Our fight to date had been for strong and gender-sensitive international legal commitments. The next stage would involve a long political struggle for these to be translated into practice. Prompted by these questions, and determined to remain part of this struggle, I made the decision to enrol in a PhD in early 2001 (Bartolomei, 2010). At the outset, my research aim had been to explore the impact of the advances in law and policy, to which the Beijing process had contributed, in improving the lives of refugee and conflict-affected women and girls. As the next stages of the story show, this very quickly changed to a research aim which instead explored why, in spite of the significant advances in law and policy, women and girls in refugee and conflict settings continued to face endemic levels of rape and sexual violence, entrenched gender inequality and pervasive barriers to leadership and participation.

Note

1. Among others these included Sunila Abeysekera and Radhika Coomaraswamy from Sri Lanka, Gita Sen and Corrine Kumar from India, and Farida Shaheed from Pakistan.

References

Australian National Committee on Refugee Women (ANCORW) (2000). *Beijing Plus Five Advocacy text*. ANCORW.

Asia Pacific Women's Watch (APWW). (2000). *Voices 2000: The little blue book*. APWW & the Australian National Council on Refugee Women.

Bartolomei, L. (2010). *Struggling against the silences: Exploring rights based responses to the rape and sexual abuse of refugee women and girls* (PhD thesis). UNSW Sydney.

Bedont, B., & Hall-Martinez, K. (1999). Ending impunity for gender crimes under the International Criminal Court. *Brown Journal of World Affairs, 6*(1), 65–85.

Bunch, C. (2000, August 22). *Summer opened with small victories, setbacks at UN*. Women's eNews. https://womensenews.org/2000/08/summer-opened-small-victories-setbacks-at-un/

Chester, B. (1992). Women and political torture: Work with refugee survivors in exile. In E. E. Collie & O. M. Rothblum (Eds.), *Refugee women and their mental health: Shattered societies, shattered lives*. Haworth Press.

Coomaraswamy, R. (1999). *A question of honour: Women, ethnicity and armed conflict*. UN Special Rapporteur on Violence Against Women. Lecture delivered at the Third Minority Rights Lecture on 25th May, 1999 at Hotel Intercontinental, Geneva, South Asia Citizens Web. Retrieved 26 March, 2008, from http://www.sacw.net/Wmov/RCoomaraswamyOnHonour.html

Copelon, R. (1994). Surfacing gender: Re-engraving crimes against women in humanitarian law. *Hastings Women's Law Journal, 5*(2), 243–266.

Department of Immigration, Multicultural and Indigenous Affairs (DIMIA). (2000). *Women at risk statistics*. Published 31 October.

Forbes-Martin, S. (1992). *Refugee women* (1st ed.). Zed Books.

Girard, F. (2000). *Further actions*. Beijing Plus Five IWHC's Analysis of Negotiations and Final Document by Senior Program Officer.

Greenlees, D. (1998, November 4). Rape terror lingers behind scars. *The Australian*. Retrieved 14 October, 2009, from www.factiva.com

Hercus, N., Ray, N., & Pittaway, E. (2000). *The women at risk program: An evaluation proposal*. UNSW Centre for Refugee Research.

Human Rights Watch. (1998). *Indonesia: The damaging debate on rapes of ethnic Chinese women*. Human Rights Watch. Retrieved October 16, 2021, from http://www.hrw.org/reports98/indonesia3/rapes.htm

Indra, D. (1999). *Engendering forced migration: Theory and practice*. Berghahn Books.

Kelley, N. (1989). *Working with refugee women: A practical guide*. United Nations High Commissioner for Refugees.

Joint Fact Finding Team (JFFT). (1998, October 23). Final report of the joint fact finding team on the 13–15 May Riot.

Mohanty, C. (1991). Under Western eyes: Feminist scholarship and colonial discourses. In C. Mohanty, A Russo, & L. Lourdes (Eds.), *Third world women and the politics of feminism*. Indiana University Press.

Mohanty, C. (1988). Under Western eyes: Feminist scholarship and colonial discourses. In R. Lewis & S. Mills (Eds.), *2003 Feminist postcolonial theory: A reader*. Routledge.

Ong, A. (1988). Colonialism and modernity: Feminist Re-presentations of women in non-western societies. In K. K. Bhavani (Ed.), *Feminism and race*. Oxford University Press.

Parpart, J. L. (1995). Deconstructing the development "expert". In M. H. Marchand & L. Parpart (Eds.), *Feminism/postmodernism/development*. Routledge.

Pittaway, E., & Bartolomei, L. (2001). Refugees, race and gender. *Refuge, 19*(6), 21–32.

Pittaway, E., & Bartolomei, L. (2005). *Only rape—Five years of activism with refugee women at risk*. Documentary film. Art Resistance, Sydney.

Salinas, M. (1988). Report on the first international consultation on refugee women, ecumenical centre, Geneva, Switzerland. *Journal of Refugee Studies, 12*(2), 292–302.

Schmuki, M. (1998). Shedding their refugee skin: Construction of women refugees and international aid regimes. *Refuge, 17*(10), 4–9.

Schrijvers, J. (1999). Fighters, victims and survivors: Constructions of ethnicity, gender and refugeeness among Tamils in Sri Lanka. *Journal of Refugee Studies, 12*(3), 307–333.

Spivak, G. C. (1988). Can the subaltern speak? In C. Nelson & L. Grossberg (Eds.), *Marxism and the interpretation of culture*. Macmillan.

Tim Relawan untuk Kemanusiaan (TRuK). (1998). *Mass rape in the recent riots: The climax of an uncivilised act in the nation's life*. Violence Against Women Division Report, Jakarta.

United Nations (1995, October 27). *Beijing declaration and platform of action, adopted at the Fourth World Conference on Women*. United Nations. Retrieved November 10, 2021, from https://www.refworld.org/docid/3dde04324.html

United Nations. (2000). *Resolution adopted by the General Assembly: Further actions and initiatives to implement the Beijing Declaration and Platform for Action*. General Assembly Resolution, 16 November, A/RES/S-23/3, United Nations, New York. Retrieved May 17, 2009, from http://www.un.org/womenwatch/daw/followup/ress233e.pdf

CHAPTER 7

Intersectionality, Identity and Refugee Women-2001–2002: 'I Will Give You a Prize If You Can Find a Woman in This Camp Who Has Not Been Raped'

Narrated by Linda Bartolomei

There were five guys who took me into the house. They raped me. Terrible, terrible things. How many times!! In the refugee camp if you want to escape you have to sleep with the soldiers or the rebels. If you don't sleep with them you will die, so what is your choice. To this day I tremble each time I go outside my door. When will it stop? (Sierra Leonean Refugee Woman, 2001).

Establishing an Activist Academic Refugee Research Centre

While gender politics were being fought at Beijing plus Five (B + 5) in New York in June 2000, a different struggle had occurred back at the University of New South Wales (UNSW). Dr Barbara Ferguson, who had established the Centre for Refugee Research (CRR) in the School of Social Work in 1999, signalled her desire to retire as the director and Eileen applied for the role. She was challenged by several committee members who felt that a more 'academic' (less activist), male professor would be more suited to the position. Fortunately, she enjoyed the

E. Pittaway and L. A. Bartolomei,
Only Rape! Human Rights and Gender Equality for Refugee Women,
Sustainable Development Goals Series,
https://doi.org/10.1007/978-981-19-0916-0_7

support of others on the committee, who recognised the value of her human rights and UN expertise and wide community links. She was appointed as Centre Director in September 2000, and a few months later, thanks to a small set-up grant, I was employed as a casual research associate.

This was an exciting, if frantically busy, time. We decided to launch CRR on the national and international stage by running two conferences, a national conference in February 2001, to follow up on the work completed at B + 5, and a large international conference in November 2001, to commemorate the 50th anniversary of the 1951 Refugee Convention. UNHCR Geneva undertook a series of global consultations to review progress and identify gaps and challenges in the protections offered by the Convention, and universities and NGO networks around the globe were encouraged to contribute their recommendations. The conferences were co-hosted with ANCORW, who following a board decision to no longer accept government funding to retain its freedom as an advocacy organisation, now shared an office with CRR. This co-location was supported by the School of Social Work at UNSW in recognition of extensive community networks and reference library that ANCORW would bring to CRR. This ensured that CRR retained a strong and central focus on refugee women and girls and opened up access to collaborative funding opportunities not traditionally available to academic institutions. Most significantly it enabled us to remain outspoken political activists donning our ANCORW hats on those occasions when speaking as academics might have constrained us!

Women Wise Up Conference: Shattering Another Silence

The *Women Wise Up* conference, held in February 2001, was organised in partnership with a range of Australian women's rights organisations. It provided an opportunity to review progress made at B + 5 and to develop new strategies to address the ongoing challenges. Many of the women from refugee backgrounds who had participated in the B + 5 Preparatory Conference were again in attendance. There was also significant involvement of women from Indigenous communities. In a workshop hosted by our friend and colleague Pam Greer, she and her Indigenous women's rights colleagues spoke of their long-held silence about the extent of rape and sexual abuse of children in remote First Nations communities across Australia and their recent decision to break that silence. Their distress was palpable, their honesty searing, but so was their terror that by speaking out they would fuel further racism against First Nations peoples and increase the number of men detained and dying in custody. While they had suffered backlash from within their communities and the broader Australian community, they, like the many refugee women who had already spoken out about their experiences of rape, sexual and gender-based violence, did not stay silent. These issues would emerge more strongly as we worked with women's rights groups across the Asia–Pacific in preparation for the World Conference Against Racism (WCAR) held in South Africa later that year.

The Theory of Intersectionality: A Turning Point in Our Theoretical Journey

In January 2001, at the invitation of Asia–Pacific Women in Law and Development (APWLD), we had travelled to Chiangmai, Thailand, to contribute to an advocacy training workshop in preparation for the WCAR. On the first afternoon, Sunila Abeysekera, one of the chief trainers and a world-renowned Sri Lankan feminist scholar and human rights activist, had introduced us to the theory of intersectionality, asking us to identify the role that racism and discrimination played in our lives and the lives of the women with whom we worked. We were a diverse group of women's human rights activists from over twenty countries across the Asia–Pacific. Some of us worked with refugee women or were refugee women themselves; others worked with Indigenous women or were themselves Indigenous women; others worked on issues ranging from trafficking, to livelihoods, to political empowerment. As women's human rights activists we were adept at identifying gender discrimination, and yet struggled to clearly explain the ways that racism intersected with and compounded gender discrimination. Sunila challenged us to pivot our focus to bring racism to the centre of our analysis, and then think about its gender impacts. It was perhaps the first time that we fully acknowledged the 'white privilege' we unconsciously enjoyed, and how discriminations experienced by the refugee women with whom we worked were the flip side of the privileges we took for granted. The conceptual framework of intersectionality, developed by black American feminist lawyer Kimberle Crenshaw (1989) had featured in discussions at the Beijing conference, and is referenced in the BPFA, but was not a framework we had yet engaged with at an academic level. It would however soon become one of our most important conceptual tools of analysis.

Teaching Intersectionality in a Football Stadium

Eileen attended the WCAR in Durban, in August 2001, and was invited to join Kimberle Crenshaw and Sunila Abeysekera presenting in a workshop to explore the concept of intersectionality with reference to refugee women and girls. They arrived to make their presentation, expecting a few hundred people in a lecture theatre. Instead, they were directed to a large football stadium hosting a crowd of over 3000 people. The crowd ranged from local villagers, grass roots activists, academics and UN and government officials. They quickly realised that the formal presentation they had planned would not work. Thinking on her feet, Kimberle suggested she use the metaphor of intersecting roads to explain the concept. 'Everyone', she had said, 'even from the smallest town, would be familiar with the idea of a main road and cross-streets'. In her presentation, she labelled each of the roads as a different discrimination, explaining that for women facing multiple discriminations, it was akin to being caught in the intersection with cars coming from all directions. Sunila took a slightly different approach, sharing stories which highlighted the way in which at different times and in different places our gender, race, ethnic, class and other identities, might expose us to multiple forms of discrimination (Fig. 7.1).

The very diverse audience loved the session, and the presenters fielded searching questions, and comments on the effectiveness of the presentation from all sectors. Inspired by the simplicity and impact of these explanations, on her return to Australia, Eileen quickly incorporated it into our Training materials. An article, *Refugees, Race and Gender: The Multiple Discriminations Against Refugee Women*, was published in *Refuge* (Pittaway & Bartolomei, 2001) and by 2017 was rated as one of their most frequently accessed articles.

Understanding the Theory of Intersectionality

The framework of intersectionality has its origins in the work of African American and third-world feminists of the 1980s and 1990s (Collins, 1989, 1991, 2000; Hooks, 1984; Mohanty, 1991). It specifically

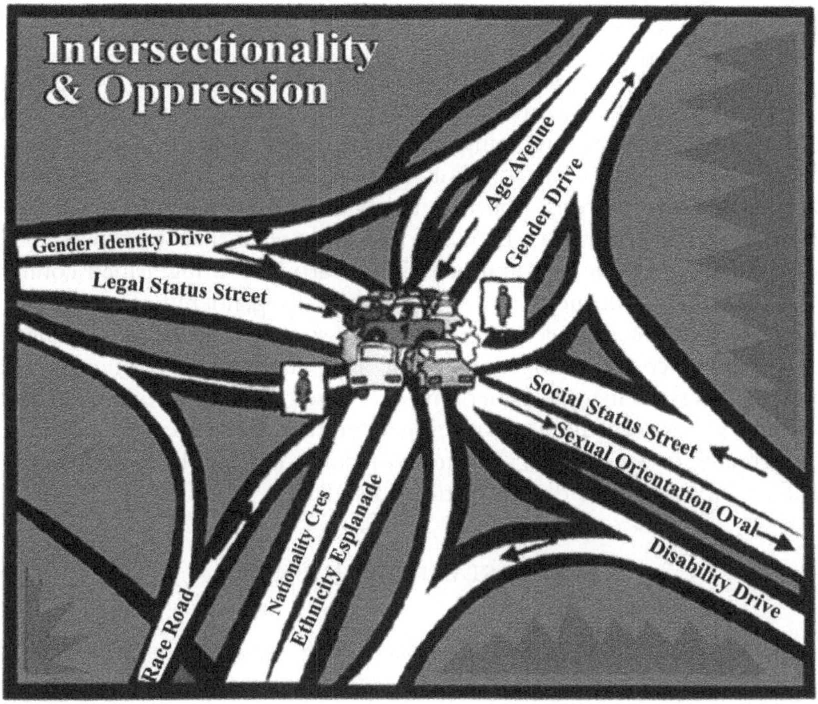

Fig. 7.1 The intersectionality road map

addresses the way in which discrimination based on race, gender, ethnicity, class and other systems of oppression can interlock to compound the disadvantage and exclusion. Feminist analysis had previously focused on the idea of multiple layers of discrimination, acknowledging that many women suffer from many different types of oppression simultaneously. Kimberle Crenshaw challenges the idea of 'layers' of oppression and discrimination, which suggests that it should be possible to separate the layers and address the issues one by one. Experience has shown that this is not effective, because it ignores the effect of being subjected to multiple discriminations. Intersectionality examines the way in which oppressions weave together, each compounding the effects of the others. The consequence is that effect of the whole is far greater than the sum of the parts.

Using Intersectionality to Analyse the Multiple Risks Faced by Refugee Women

To use this model as an analytical tool, we had to name and explore the origins of the oppressions refugee women face, and the impact of these. They are the result of racial, ethnic and religious conflicts; the aftermath of colonisation; and racist policies and law. Refugee women suffer discrimination on the grounds of their sex and because of their membership in a marginalised ethnic, cultural or religious group. During armed conflict and the refugee journey, they suffer specifically gendered atrocities that are not shared by refugee men in the same social group. In resettlement, they experience racism on account of their ethnicity or religion, and a lack of access to income generating and resettlement opportunities due to their gender. In the lives of refugee women, these are not individual occurrences or events with separate root causes that can be traced back to one or another of the discriminations they face.

Understanding Identity

> *An intersectional analysis [of oppression] involves an analysis of a situation from a perspective based on the understanding that we all have shifting and multiple identities.* (Sunila Abeysekera, 2002)

Building on this analysis, recognising the importance of identity became another key tool to help us understand the impact of intersectional oppression on refugee women and girls. We drew on a wide body of theory to support the analysis of individual identity as something that is not fixed or static (Allen, 1998, p. 50). Depending on the particular social circumstances at a given time, one or another identity is brought to the fore: one's primary identity therefore changes according to the social environment.

For example, a refugee woman is often seen by the world through the lens of international media as a pitiful and vulnerable, a victim of violence, in need of food and protection. However, each refugee woman also carries a number of different identities which become relevant at different times and in different situations. A refugee woman has gender, race, nationality, ethnicity, religion, a socio-economic and a socio legal status, all of which are socio-political identities. The experiences endured by refugee women are not caused by discrimination against one, or some, or even all of these identities, but by the oppression caused by the compounded effects of discrimination against them all. To fully understand the experiences suffered by refugee women, it is necessary to analyse how intersectional oppression against their multiple identities impacts on their lives and their ability to access human rights (Real, et al., 2002). The graphics below demonstrate the impact that these different political labels have on the way in which refugee women are seen and treated.

A Refugee Woman's Identity Labels

The first refugee woman is wearing identity labels given to her as a result of intersectional oppression. She is a vulnerable, impoverished, anonymous refugee woman, such as you might see in news reports of an armed conflict (Fig. 7.2).

The second refugee woman is strong, proud and resilient. She is no longer anonymous, she is part of a family and a community. She has agency over her life and is someone we admire (Fig. 7.3).

These two women are in fact the same person, and as can be seen in graphic three, she wears all of these labels (Fig. 7.4).

Effective 'protection' for refugee women means addressing the negative labels, while enabling the positive.

Refugee Woman: A Dangerous Label

Discovering and applying the theory of intersectionality was another 'lightbulb' moment in understanding both why refugee women and girls were at greater risk of rape and sexual violence than other women, and why, despite the advances in international law and policy, this issue was still not addressed. One answer is because she is wearing the label of 'refugee woman' with which she is consistently branded. This imposed identity defines refugee women as an extremely vulnerable and exploitable group: while the term 'woman' denotes a person of low sociocultural status, the term 'refugee' conveys their corresponding lack of legal-political status. The intersection and compounding effect of these identities as both women and refugees can explain their extreme vulnerability to rape, sexual and gender-based violence, by actors who are confident of legal impunity and social acceptance.

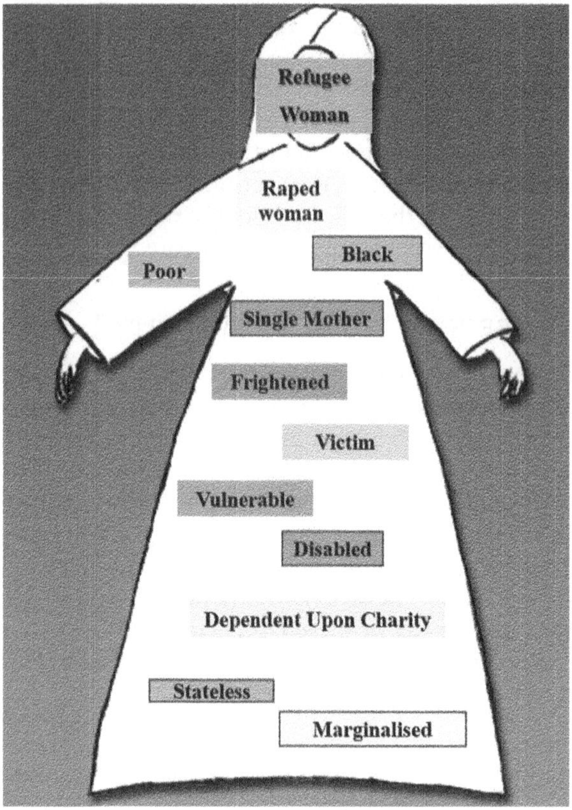

Fig. 7.2 Refugee woman with negative labels

While all women are vulnerable to rape, it was becoming increasingly clear that not only was the risk for some women, including refugee women, far greater than for others, but that this increased risk was often directly correlated with a decreased likelihood of prosecution of the perpetrators. This was not just simply a question of impunity but of the politics of impunity. Ending impunity for the perpetrators of violence against women had characterised the efforts of feminists throughout the 1990s. While the call might have been easy to make, the complexity of achieving this was clearly far greater than recognition in law. Questions of class, race, location and power all needed to be interrogated.

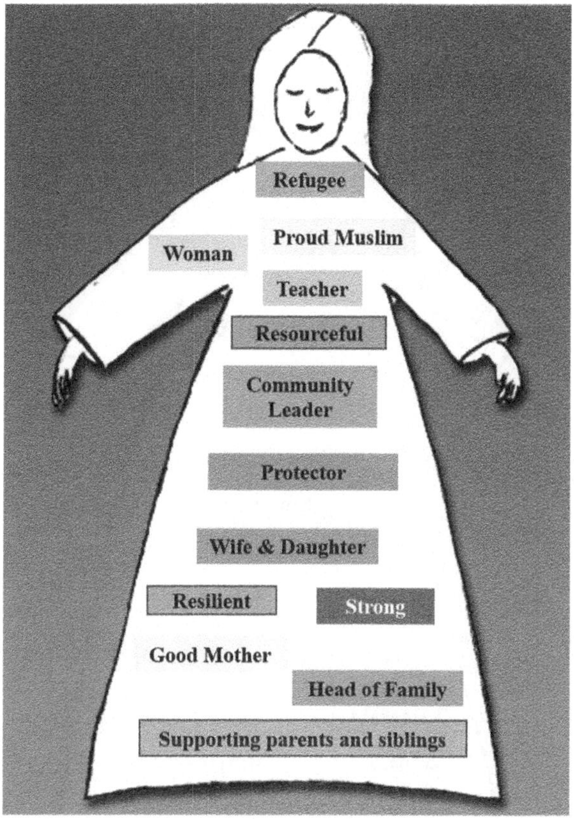

Fig. 7.3 Refugee woman with positive labels

In her article exploring why the international community had mobilised so strongly around the white women who had been raped by white men in the Former Yugoslavia, Rhonda Copelan asked these questions. She highlighted the power of racial identity and privilege as one of the answers: white women are valued more highly than women from other ethnicities (Copelon, 1994). As we have already discussed, when similar atrocities happened to the women of Rwanda, the world only learned about it much later, and similarly barely anything was heard about the Chinese women raped and sexually tortured in Indonesia in 1998. The

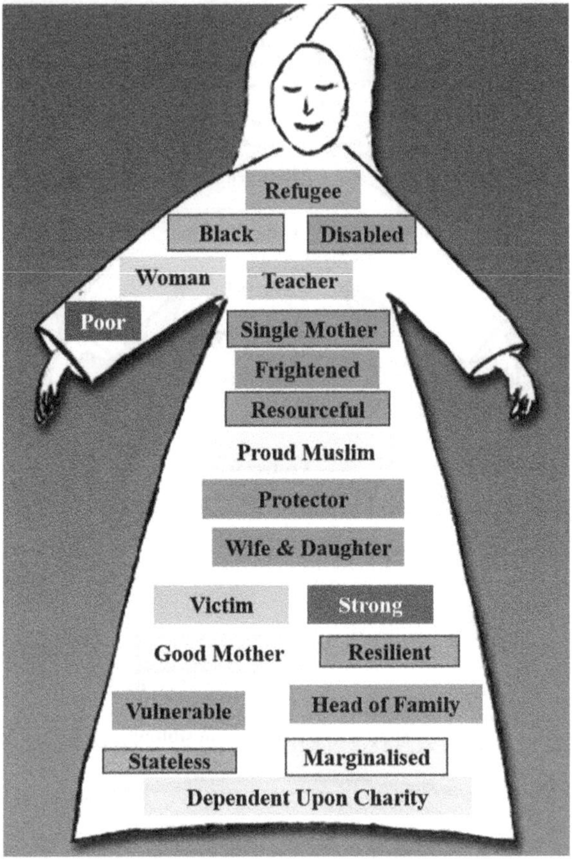

Fig. 7.4 Refugee woman with positive and negative labels

international media had focused almost exclusively on the ethnic dimensions of the Rwandan genocide and the atrocities perpetrated against the ethnic Chinese community in Indonesia, rather than on the rape and sexual torture. The implication was that when such atrocities occurred in non-white communities, it was 'ethnic tension', whereas when they were perpetrated against white women, it was an outrage. While at a surface level we were aware of this discrepancy, only now did we begin to fully apply this analysis in our advocacy.

Struggling for Refugee women's Rights Against a Backdrop of World Changing Events

World events of the time had a significant impact on the struggle to retain an international focus on the rights of refugee women and girls. During this period, protracted civil and interstate wars continued to rage across Africa, including in the Democratic Republic of the Congo (DRC), Sierra Leone, Angola, Liberia, Algeria and Chad. Millions of people died, either killed in the conflicts or from starvation and disease. Millions were internally displaced or sought asylum in neighbouring countries. The mainstream media in Australia and across countries of the global north provided at best limited and intermittent coverage of these conflicts. However, in limited segments of the media we noticed a discernible shift as the rape and sexual abuse of women and girls in the conflicts, particularly in the DRC, was finally being reported. This shift in media reports followed by the widely publicised prosecution of rape in the International Court of Truth and Reconciliation (ICTR), and the formal establishment of the International Criminal Court (ICC) in, July 2002, marked positive progress.

Three days after the conclusion of the WCAR, the world was shaken by the September 11 terrorist attacks on the World Trade Centre in New York. The response of the USA and its allies, to launch aggressive military attacks, first against Afghanistan and then Iraq, contributed to an exponential growth in racism and Islamophobia, igniting extremist Islamist movements. Across Europe, the USA, Canada and Australia, it fuelled popularist anti-refugee and anti-migration political movements and propelled previously marginal xenophobic voices to the centre. All of this had direct and immediate consequences for refugee women and girls, and refugee policies in countries of asylum, both rich and poor, became increasingly restrictive. In Australia, a deeply political battle between refugee rights and national border sovereignty dominated the political and media landscape throughout this year, with devastating consequences for the human rights of those fleeing persecution.

Demonising Refugees

[W]e will decide who comes to this country and the circumstances in which they come. (John Howard, Election Policy Speech, October 28, 2001)

During this period Australia had begun to increase its intake of refugees from Africa as part of the resettlement program, including from Sierra Leone and Sudan, with a small number arriving on Women at Risk (WaR) Visas. However, in the election year of 2001, this positive progress was completely overshadowed by increasingly restrictive asylum policy, which would leave Australia regarded as a pariah on the world stage. Since 1999, in response to an increase in the number of asylum seekers arriving in Australia by boat, the Government had developed a series of increasingly draconian deterrence measures. First, Temporary Protection Visas (TPVs) were introduced. These visas put a time limit on the protection provided to boat arrivals and did not allow family reunion. As many men had come ahead of their wives, this had particularly negative consequences for refugee women and children who were often left stranded in transit countries and countries of first asylum. More families risked dangerous journeys to seek asylum and we increasingly heard of rapes at sea by smugglers, and many women and children drowned. In August 2001, the Tampa, a Norwegian freighter, rescued 433 asylum seekers, from their sinking boat in the middle of the Indian Ocean. As the asylum seekers were on their way to Australia, the ship's captain sought permission to bring them here, but permission was denied, leading to a stand-off on the open seas. The asylum seekers were eventually transferred to an Australian navy vessel and taken to the Pacific Island state of Nauru. This led to the Australian Government passing a series of migration bills, with bipartisan support, which became known as the Pacific Solution. These laws institutionalised the protracted imprisonment of refugees on remote islands across the Pacific, and if not in direct breach of international refugee law, they directly breached its spirit and marked Australia as a nation willing to place political self-interest above human rights.

This was followed by what became known as the 'children overboard affair'. On October 7th, in the lead up to the 2001 federal election, several government ministers claimed that asylum seekers on board a boat which the navy was attempting to turn away from Australian waters had thrown their children overboard in a ploy to secure rescue and passage to Australia. These claims were subsequently disproven in a Senate Select Committee inquiry, and the Government was criticised for misleading the public and fuelling voter fears of mass arrivals (Phillips & Spinks, 2013). The Government, seeking to defend its restrictive asylum policy, frequently demonised asylum seekers as 'queue jumpers' and, particularly those coming from the Middle East, as terrorists (Adams, 2001).

They implied that refugee camps in countries of first asylum were places of safety and that by fleeing them refugees were trying to jump the 'queue' ahead of other and perhaps more deserving cases. This triggered a backlash felt most strongly by refugees from Muslim backgrounds living in Australia. They faced increasing racism and Islamophobia, with deeply gendered and intersecting consequences for visibly Muslim refugee women. Being identified as a Muslim refugee woman in Australia was indeed a *dangerous label*. We were not alone in feeling deeply ashamed of what our Government did in the name of the Australian people.

The Truth Hurts

Following the Tampa incident, racist abuse towards refugees from a Muslim background increased exponentially. The refugee women working with us reported the extreme distress suffered by their communities. As women wearing hijabs, they frequently experienced verbal abuse, were spat on and had their scarves torn from their heads. These were the days before social media. In an effort to reach the community, CRR and ANCORW, produced a booklet, *The Truth Hurts*,[i] which addressed the myths surrounding refugees which were fuelling the xenophobia. It was intended for use by refugee activists and service providers. We had a small print run of 700, which we launched at the *Where to From Here?* conference in November 2001. It became such an important advocacy document that we sold over 40,000 copies over the next two years, and even though it was focused on Australia, other resettlement countries bought, amended and distributed it.

Where to from Here? Refugee Conference Sells Out—And We Dance!

The refugee conference, co-hosted with a network of Australian civil society organisations in November 2001, was an enormous success. In our early planning we had optimistically anticipated registrations of 200, providing discounts for small organisations and fully funded places for refugees. Little did we know that the Australian Government's response to the Tampa and the 'children overboard affair' would mobilise NGOs, activists and academics across Australia. The timing of our conference meant that it was one of the first formal opportunities for all to come together. Days out from the event, registrations skyrocketed from 200 to

800. Without the help of students, family and friends jumping in to assist we would never have coped and even then, we had to beg forgiveness for crowded rooms, overflowing sessions and misprints in programs.

To ensure a central focus on refugee women and girls, we hosted a one-day Refugee and Indigenous Women's Human Rights Tribunal[ii] the day before the conference. The Tribunal was underpinned by a southern feminist approach privileging women's voice and agency as partners in the struggle. Refugee and Indigenous women shared testimony of their experiences of rape, sexual and gender-based violence, due to conflict and colonisation. The fear of a racist response was raised as one of the many factors that kept both refugee and Indigenous women silent about the rape and sexual abuse they faced within their own communities. Many of the women also joined the more formal academic conference in the days that followed. The strength of their narratives added a powerful human dimension ensuring the prominence of refugee voices, powerfully drawn from their lived experience and highlighting their wisdom, knowledge and analysis.

Who Am I?
I am a refugee. I am... a dirty woman, hopeless, a hungry person, an ignorant person, a troublesome person, yet again another burden for the world to feed, another burden for the world to care, that is who we are... we have been labelled. But being a refugee is not by choice. If it is a choice, I wouldn't be a refugee anywhere. Looking at myself, I believe I am not a victim, but I am a survivor, a very strong person, a refugee woman (Sudanese refugee woman, 2001 in Bartolomei & Pittaway, 2003).

This conference was the antithesis of the dry academic conferences with which so many of our colleagues were familiar. Fatmata Sesay, an incredibly charismatic guest presenter from Sierra Leone, a refugee and an academic living in Belgium, ended her keynote address on the first day with an exhortation that we all should get to our feet and dance. She explained that in her culture, singing and dancing had a deep significance, and that in joining together it would break down barriers and enable the sharing of experiences and ideas. Etched in our memories is the image of older white men in suits, hips swinging and ties flying. The mood of the three-day conference was set, the ice was broken, the atmosphere was warm and respectful, and the discussion honest and fruitful. Another dancing moment!

People commented very positively about how unusual it was to have so many women and men from refugee backgrounds presenting in sessions: 'very different from the usual academic conferences' was echoed time and time again. We were pleased with this response, but it also highlighted for us how far we had to go, because in fact presenters and participants from refugee backgrounds were still very much in the minority. With the support of an incredible team of NGO and student volunteers, every conference session was fully documented, and each night the key outcomes were synthesised. In the final conference session nominated rapporteurs[iii] from each stream presented the key outcomes and recommendations. The summary conference report and our major recommendations were then emailed to an NGO colleague in Geneva the following morning so that they could be formally tabled along with the outcomes of other regional conferences and events that had been held across the globe as part of the 50th Anniversary Consultation process. Academics and NGOs also contributed recommendations from consultations held with refugee women and girls.

50th Anniversary of the Refugee Convention: Dialogues with Refugee Women

As our advocacy efforts for refugee women and girls during 2000 and 2001 had still been principally focused on the Beijing process, we had not actively engaged with UNHCR. In partnership with the Women's Commission for Refugee Women and Children (WCRWC),[1] led by the new senior coordinator for refugee women/gender equality, Joyce Mends-Cole, UNHCR had conducted a series of local and regional meetings with refugee women in over 20 international sites. This process culminated in a ground-breaking Women's Dialogue in Geneva, in June 2001. It was the first time that refugee women had been sponsored to attend a formal meeting of UNHCR in Geneva. The Dialogue was attended by 50 refugee women from refugee sites across the world. The testimonies shared by the refugee women clearly illustrated the enormous

[1] Now Womens Refugee Commission (WRC).

gap between UNHCR's policy commitments and the on the ground realities. The summary of recommendations from this meeting identified that the issues of concern raised by the refugee women were '***mostly not new ones***' and emphasised that progress made was far from even (UNHCR, 2001, p. 10). In other words, the main problems, as identified by the refugee women, were failures in implementation. The most important outcome of these 2001 Dialogues was that the High Commissioner made five commitments to refugee women and girls.

The Original Five Commitments to Refugee Women 2001

1. Develop integrated country-level strategies to address violence against refugee women, including in situations of domestic violence
2. Refugee women will be registered individually and delivered relevant documentation to ensure their individual security, freedom of movement and access to essential services.
3. Refugee women and men will be enabled to participate equally in the registration process. In all management committees and other mechanisms of representation of refugees to UNHCR in urban, rural and camp settings, including return areas, 50% of representatives will be women.
4. Refugee women's direct and indirect participation in the management of distribution processes for food and non-food items will ensure, to the extent possible, that these are distributed and controlled directly by adult female household members.
5. The provision of sanitary materials to all women and girls of concern should become a standardised practice in UNHCR assistance programs.

Despite this success, fear of refugees coming to participate in meetings in Geneva absconding and seeking asylum in Switzerland was so great that even though it did not happen, the invitation was not extended again for a further ten years.

Meanwhile Australia's Women at Risk Program Continues to Fail

In early 2000, we received reports that the WaR Program was again struggling to fulfil its potential, failing to meet its allocated quota of 10.5% of Australia's resettlement places. Australian Immigration (DIMIA) and

UNHCR staff had also expressed concerns that the Program was not targeting those most at risk. In 2001, in partnership with ANCORW, we secured a small grant to continue research into the ongoing challenges in the implementation of the Program and identified two sites in which to undertake the work. As the Program is a global program it was essential that contrasting sites be selected in order to explore how a single category of risk might apply across diverse settings. Kenya was selected as one of the sites, since Australia's Nairobi post was the main hub in Africa for processing WaR applications. The second site selected was Pakistan, which at the time was struggling to process a large number of applications from Afghan refugee women amidst increasing accusations of fraud (Hercus et al., 2000). Because access to refugee camps to conduct research generally required formal permission from either UNHCR or the relevant camp authorities, connections with UNHCR and NGO networks on the ground were crucial to us securing access.

Critical Connections and Networks

As the project funding was limited, the only way to afford to visit both Kenya and Pakistan was by adding the travel to already planned and funded overseas trips. Travel to Kenya and access to Kakuma refugee camp was made possible by a short consultancy, from a Sydney-based aid organisation AUSTCARE, to evaluate a program they were funding in Kakuma refugee camp. Eileen planned to undertake the research in Pakistan on her way home from the WCAR in Durban in September 2001. She had departed Durban, spent a few days in London but as she was about to board the flight to Islamabad, news of the September 11 attacks on the World Trade Centre in New York broke. Very quickly responsibility was laid at the feet of al-Qaeda, and travel to Pakistan was off the cards.

Unable to travel to Pakistan, an alternative site in Asia needed to be identified. We chose Thailand, on the basis that it would also provide a strong contrast with Kenya, as it was not an identified WaR processing hub. But most importantly, we had established links with several community-based refugee women's organisations who, if they agreed to partner with us, could assist us in accessing the refugee camps. In the training workshop we had held in Chiangmai the previous year there were representatives from the newly formed Women's League of Burma (WLB), a network of 13 refugee women's organisations from different ethnic groups, which had held its inaugural congress meeting in January 2001. Senior members shared their aspirations and outlined ways in which the international women's movement might support them. One woman told of how tired they were of attending meetings of the UN only to be allocated a non-speaking support role by the men in the Burmese democracy movement. They were told 'to look pretty, wear their ethnic dress and most importantly to stay silent'. Working with this group of refugee women deepened our understanding of the many intersecting and multi-layered barriers that refugee women faced in their quests for safety, justice and gender equality. We were delighted when they agreed to collaborate with us.

Seeking Safety, Finding Terror

In January 2002, we travelled to Thailand and then to Kenya. In each site, meetings were held with a range of NGO and refugee women to explore how the WaR Program was being implemented at the local level. The aim of these meetings was twofold: one was to build a preliminary understanding of the protection environment for refugee women and girls in each site; the second was to assess the extent of local knowledge, understanding and implementation of the WaR Program.

Refugee Women at Risk in Burma and Thailand

In Thailand, we first travelled to the Thai–Burma border town of Mae Sot and visited the Mao Tao Clinic run by Karen/Burmese refugee, Dr Cynthia Maung. This clinic provided free health care to Mae Sot's large urban refugee population and also supported a series of cross-border health services. We met with a group of young women working as 'backpack medics' in difficult and dangerous conditions inside Burma. They shared accounts of the trauma and serious mental health problems faced by women who had been raped and sexually tortured, and of their inability to provide support or counselling. We saw the documented accounts they were collecting of thousands of women who were 'recruited' at gunpoint as porters for the military and then raped (Apple, 1998, p. 18). We heard that many gave birth to babies of rape or died from self-induced abortion. Others suicided from an overwhelming sense of shame.

> *One of the founding members of WLB recounted a story of a journalist travelling into Shan state in Burma. He had stopped one day at tea stall by the side on the road in a small village. Looking up, he saw a row of tattered ropes swinging in the wind, and asked the shopkeeper what they were for. He responded that they were the ropes used by the many women in the village who had been raped by the military to hang themselves* (Pittaway & Bartolomei, 2003).

A Complete Lack of Protection for Refugee Women

With the support of WLB, we interviewed refugee women leaders and NGO staff who were working in the refugee camps along the Thai–Burma border. All described their struggles to provide protection to women who had suffered rape, sexual and gender-based violence, in Burma and in Thailand. While some of the camp-based refugee women's groups had established informal safe houses, they had no access to counselling services. None had even heard of the WaR Program. In the previous six months many women, forced to work as porters in Burma, had fled to the Thai border camps. However, none had yet been able to be registered by UNHCR. The evidence they had begun collecting demonstrated that the sexual torture of women and girls was part of the Burmese military regime's strategy of power and control. They also emphasised that the risks did not cease once the women reached Thailand. Those who managed to flee Burma and cross the border often then faced rape by border guards. Those who were entitled to enter the camps found no counselling and limited support. Others, who were denied entry to the camps, remained undocumented and forced to work illegally, risking rape by employers or by Thai police during deportation (SWAN, 1999, 2001).

Rape Is Still Not Considered as Grounds for Refugee Status

When asked about why refugee women at extreme risk in Thailand were not referred for resettlement under the WaR Program, UNHCR staff advised us that it was due to the Royal Thai Government's reluctance to issue exit visas. They feared that a large resettlement program might act as a 'pull factor' for more refugees to cross the border from Burma. The exception to this was for those refugees who had been involved in the 1988 student uprising in Burma. These mainly male refugees were perceived as being too political, and the Thai Government was happy to have them resettled elsewhere. Despite the commitments made at Beijing and again at B + 5 to consider gender-related persecution as grounds for refugee status, in practice, political asylum and resettlement favoured men, as refugee status determination still drew on traditional definitions of torture and political activism. Rape and sexual violence were still regarded as private matters and consequently, the rape of women was not considered political. Women who were raped were not recognised as having been tortured and were not granted refugee status. Rape may have been recognised as torture and a war crime in the Statute of the ICC and been prosecuted in the International Criminal Tribunals for the former Yugoslavia and Rwanda, but as yet this had had no discernible impact on the situation of the many hundreds of women who had fled Burma following rape by the military.

This realisation was deeply disturbing. The risks of rape and sexual violence were so much higher and the response mechanisms so much weaker than we could possibly have imagined. What we learned in Thailand shattered any illusions we may have held about the equitable implementation of UNHCR protection standards and the global reach of the WaR Program.

And then to Kenya

Going from Thailand for Kenya, we comforted ourselves that even if more problems were discovered, we knew that the WaR Program was operational there. We were utterly unprepared for what we would discover in Kakuma refugee camp. The sickening sense of horror and desperation we felt as we arrived, will never go away. It followed a flight from Nairobi to a remote desert airfield from which mercenary pilots flew airdrops of food into the South of Sudan. This was followed by a hundred kilometre drive along a dangerous road in an armed convey. Kakuma is a camp of 95,000 refugees, one hundred kilometres from the Sudanese border, in the harshest, most isolated part of Kenya. Year round the average daytime temperature is 40 C, dropping only slightly at night. The terrain is flat, barren and dry. Dust storms are an almost daily occurrence, water is scarce, and rain is occasional and then leads to flooding.

Vegetation is sparse and in many parts of the camp non-existent. Cultivation of crops is almost impossible due both to poor soil quality and the lack of water. The day we drove into the camp it was during one of the all-too-common dust storms, 40 degrees in the shade and the air was thick with blinding, red, gritty dust. We felt nothing short of horror as we peered through the dusty fog at rows of white plastic tents in which a third of the camp's population had lived since floods had washed away their mud huts some eight months earlier. Our distress increased when we saw the camp hospital which was still in a tent, lacking medications or bedding. Then the 'Protection Area' which was a barbed wire enclosure where single women and families at extreme risk were housed in filthy canvas tents for their own safety. Many had been confined there for seven years often following multiple and brutal rapes, attacks and the murder of family members. UNHCR and NGO staff reported that the incidence of rape and sexual violence was endemic in the camp.

Endemic Rape, Sexual and Gender-Based Violence
The abduction and sale of young girls as brides, the forced marriage of widows, the physical and sexual abuse of those in mixed marriages and domestic violence were reportedly commonplace. There were also many reports of gang rapes, and sexual mutilation by gangs of men. Women who were raped by rival groups and gave birth to babies of rape, were stigmatised and harassed, and were in urgent need of protection. Yet for most of these women there was simply no protection available (Pittaway & Bartolomei, Fieldnotes, 2002a).

The women cited a number of factors which contributed to this situation. These included: conflict between clan groups and with the local Turkana people; a high rate of alcoholism; lack of economic independence; and an almost complete lack of social structure inside the camp. Kakuma was one of an increasing number of so-called protracted refugee settings around the world, in which refugees had been forced to remain effectively 'imprisoned' and forgotten for years and even decades, in isolation, danger and terror. The devastating impacts of protracted encampment on mental health and on the social fabric had already been well-documented (Africa Watch, 1993; Crisp, 2000; Harrell-Bond, 2002; Hyndman, 2000; Kelley, 1989). This visit exposed us to the visceral reality of those impacts and the way in which encampment violated not only the civil and political rights of the refugees to freedom of movement, legal protection and justice, but also denied their economic, social and cultural rights (Harrell-Bond, 2002). As became clear on the second day of our visit, not only were most refugees in Kakuma unable to engage in sustainable livelihoods, most did not have enough to eat.

Astounded and Stunned

We had set out to explore why, given what we knew about the number of women at risk in refugee situations, the WaR Program was still failing to meet its potential, but we were not expecting failures of this magnitude. Shortly after arriving in Kakuma we met with the UNHCR Head of Camp. He listed the many and multiple risks faced by women and girls and agreed that resettlement was an important solution, but noted that with very few places available, providing education and viable livelihoods in the camp were essential protection measures. The direct link between the lack of livelihoods, subsequent poverty and increased risks of rape, sexual and gender-based violence, were obvious. Importantly he noted that while there were numerous small-scale so-called income generation projects in the camps, most were utterly useless as there was no market for the goods produced. When we asked about WaR, he swept his hand towards the window and the camp outside and said: 'They are all at risk – are you going to resettle them all'? (Pittaway & Bartolomei, Fieldnotes, 2002a).

'Welcome to Kakuma': The Story that Had to Be Told

We resolved during this visit to return to Kakuma to document the appalling human rights abuses and problems faced by the women and their families. At the request of members of the refugee communities, NGOs and UN agencies, we also undertook to advocate that additional resources and resettlement places be directed to Kakuma. A key advocacy tool we produced from this visit was a short four-part film entitled *Welcome to Kakuma*. The choice of title is clearly facetious, because Kakuma was so obviously a place to which no-one would wish to be welcomed, yet we were frequently greeted with these words while we were there. In hindsight, this seemed less disturbing than being told to '*have a good time in Kakuma*' by numerous people during stopovers in Nairobi on subsequent visits. It starkly emphasised the need to raise awareness about the realities of camp life.

Making a film had not been on the agenda and we had only small video camera with us. However, the refugees all begged us to take photographs of them, to film the conditions and to let the international community know how they were suffering. The short film provided a dramatic picture of the camp, depicting the haze of heat and dust swirling around hundreds of plastic tents. It described the over-crowding, the lack of food and water, the terrible risks of violence, rape and sexual abuse and complete lack of any kind of protection.

One section of the film focused on a small group of refugee women. Two had been raped during the conflict in Sudan and then again in Kakuma camp. Both had borne babies from these rapes and one was pregnant a second time from another rape. Though each of them gave permission for their photographs to be taken, we decided that because there was a strong possibility that they might later be resettled to Australia, we blurred their faces and those of their children. Our decision proved wise, as both women and their children did eventually come to Australia.

Elizabeth's Story

One of the women we met and filmed in Kakuma, subsequently resettled to Australia, would join us as a presenter at a conference held by the Centre for Refugee Research in December 2005. She approached us and asked us to film her presentation and to then take '*the marks off her face*' in the Kakuma film and edit the two clips together. She said she wanted the world to see how she had been in Kakuma and how well she was doing here. Her story is shared in a ten-minute DVD entitled *Elizabeth*.

Welcome to Kakuma later proved to be one of the most powerful documentary and advocacy tools we ever produced. It was shown at UNHCR Geneva in October 2003 and contributed significantly to advancing our advocacy for an improved response to WaR by both UNHCR and member states. It also sold hundreds of copies across Australia and overseas and was much accessed on YouTube. As a result, our cameras became our regular research companions and videos became a component of all future research reports and advocacy campaigns.

Myths, White Washing and 'Royal Tours'

What we witnessed in Kakuma directly challenged the myths that were circulating at the time in the countries of the global north. These included that refugee camps were places in which refugees enjoyed safe and effective protection; and that those refugees who managed to get out and make their way on horrendous journeys to countries in Europe or Australia were 'queue jumpers' who were 'by-passing' the effective protection offered in countries of first asylum. For the refugees in Kakuma, there was no effective protection – in fact for many, there was no protection at all. There also was no functional 'queue'. The then Australian Immigration Minister, Phillip Ruddock, had visited Kakuma camp in the late 1990s and frequently used the trope of the 'vulnerable raped' women in Kakuma to demonise Middle Eastern asylum seekers arriving in Australia by boat, implying that Middle Eastern asylum seekers were 'taking the places' that should be open to the women from Kakuma (Radio National, 2002). As we had seen first-hand, however, those places weren't being offered to the women in Kakuma. Given Australia's continued failure to fulfil even the very small number of places allocated to women at risk, this was yet another example of the divisive refugee politics gripping Australia. Moreover, during this first visit to Kakuma, we had heard from NGO and UNHCR staff, that 'our Minister' had not even seen the real Kakuma but a literally 'white-washed' version, on a visit akin to a 'royal tour'. Care had been taken to ensure that all of the project offices and buildings he had visited had been freshly painted. Medicine cabinets in the camp hospital had been temporarily stocked and clean sheets put on the hospital beds—all quickly removed once he had left. We would subsequently see similar 'royal tours' by government officials in refugee camps across the world. Politicians and UN officials were rarely exposed to the realities of life in refugee camps and many humanitarian staff on the ground were complicit in maintaining the myth that refugee camps were places of safety and protection.

Our final day in Kakuma highlighted the terrifying reality of life there. We met with the UNHCR Head of Camp again at an awards ceremony which Don Bosco, an INGO, which provided livelihood training, was hosting for their students. Both the Head of Camp and Eileen had been invited to provide 'motivational' speeches. While looking out at the crowd of refugee students and their parents, he had turned to Eileen and said,

'I will give you a prize if you can find a woman in this camp who has not been raped'.

And then Came the News from West Africa...

Within days of returning to Sydney, news broke of research in West Africa undertaken by UNHCR and Save the Children revealing disturbing and endemic sexual abuse of refugee children by aid workers and peacekeepers. The report included allegations that at least 40 children, mainly girls, had been raped and sexually abused. 'Young girls reported exchanging sex for desperately-needed humanitarian assistance – biscuits, soap, medicines – or meagre sums of money' (Asmita Naik, 2003, p. 13).

> *I leave my child with my little sister, who is ten years old, and I dress good and I go where the NGO workers drink or live and one of them will ask me for sex, sometimes they give me things like food, oil, soap and I will sell them and get money.* (Refugee girl child)
>
> *They change girls so much and none of them marry the girls - if she becomes pregnant she is abandoned, with no support for herself and the child. Most of us used to just look at them and wonder. Our brothers, they have a problem.* (Aid worker)
>
> Victims' voices: testimonies of abuse from West Africa, 2002 in Asmita Naik, 2003.

This was widely reported in media around the globe and prompted an outcry. However, the response from the international humanitarian community was mixed. On the one hand, governments took to the floor in meetings of UNHCR's Executive Committee to demand action, urgent meetings were held and an interagency taskforce was established. On the other hand, attempts were made to discredit the veracity of the findings. Most disturbing was the response of the then UNHCR High Commissioner, Ruud Lubbers. In a CNN interview in May 2002, he stated that 'we hardly find concrete evidence. It is very scarce' and doubted that the cases described in the report constituted exploitation at all: '(the) mother is only happy when it happens, because it is one person less to feed'; 'mothers are just delighted when [their daughters] can find a husband'. The interviewer replied, 'There's no talk from the girls or UNHCR's own report of anything even remotely approaching romance' (Asmita Naik, 2003, p. 13).

Many UNHCR staff were appalled by these claims and a strong memo criticising his comments was sent to him by the UNHCR Staff Association. Given the clear misogyny of his remarks it will perhaps not come as a surprise that in 2005 he was forced to resign his position as UNHCR High Commissioner, following allegations that he had engaged in a pattern of sexual harassment of junior female staff (The Guardian, 2005). This was just one of many similar incidents we would encounter, as time and time again, powerful and misogynistic men sought to use their institutional power to hide and deny their own complicity in the rape and sexual abuse of refugee women and girls.

Moving Forward—Lessons Learned in This Research and Advocacy Cycle

> Although women's rights are now recognised as human rights, recognition does not mean implementation. Much work still needs to be done to achieve human rights for all. (Fraser, 1999, p. 906)

These fieldtrips fundamentally challenged our faith in the ability of law and policy to directly inform practice. We had set out on this next stage of the journey to explore how the shattering of the silences, and advances in law and policy relating to recognising and addressing rape, sexual and gender-based violence, were impacting on the lives of refugee women. Still naïve, we genuinely had not expected such little progress. The question now became not what had changed, but why so little had? It was clear that changes in law and policy were not the only strategies required if refugee women and girls were to enjoy their human rights to bodily integrity. We drew again on the critical policy and discourse theory to try to make sense of the political and ideological factors which continued to deny refugee women at risk recognition and access to safety and protection. In Thailand, political factors limited UNHCR's role and denied refugee women at risk access to resettlement and to other forms of effective protection. In Kenya, there seemed to be a level of fatalism with regard to the level of risk for women and the limited capacity of UNHCR to respond.

In both Thailand and Kenya, we had identified refugee women and girls who clearly met the WaR criteria (UNHCR, 1998) and yet none of them had been considered for resettlement under that Program. In both sites, problems in the implementation had been identified. We submitted a report at the ExCom meeting in Geneva in October 2002 and requested that we work more closely with UNHCR to seek solutions. This marked the beginning of an ongoing research collaboration with the organisation (Pittaway & Bartolomei, 2002b). Based on the findings from these preliminary field trips, in partnership with ANCORW, we applied for, and received further Australian Research Council[iv] (ARC) funding to explore the apparent failure of the WaR Program.

Next Steps

The main aim of the ARC-funded research was to clarify the barriers to effective implementation of the WaR Program and to develop strategies to overcome these. Having principally focused our research and advocacy efforts on influencing the UN Commission for the Status of Women (CSW) through the Beijing process, we now began to engage more actively in formal international meetings convened by UNHCR in Geneva. We wanted to know whether the law and policy advances achieved through the Beijing process were having any substantive influence on UNHCR's international policy frameworks. Our policy advocacy began increasingly to feel like a double helix as we worked to weave together the two interlinked strands of our work in the field and our advocacy at the UN. We juggled teaching and university commitments, and in my case a very part-time PhD, with regional women's conferences and UN meetings at both CSW in New York and at UNHCR in Geneva, while our fieldwork in refugee camps and urban settings increased exponentially.

Notes

1. Belinda Lucas generously provided voluntary research assistance.
2. Without the work of ANCORW Board Members Kiri Hata and Elenie Poulos and UNSW academic colleague Sue Green, this event would not have been possible.
3. Conference rapporteurs included, Caroline Lambert, Sue Harris-Rimmer, Alister Gee and Graham Thom.
4. We were joined by NGO partners; ANCORW, AWHRC, Public Health Association of Australia (PHA), The Jesuit Refugee Service, the Refugee

Consortium of Kenya, academic colleagues Anna Whelan, Anthony Zwi and research assistant Deborah Raphael.

References

Abeysekera, S. (2002). Introduction: Intersectionality. In M. J. Real, A. Aggarawal & J. Pasimio (Eds.), *Intersectionality of women's rights*. APWLD.

Adams, P. (2001, September 18). *Are refugees terrorists? Late night live broadcast*. ABC Radio National. Retrieved November 22, 2021, from https://www.abc.net.au/radionational/programs/latenightlive/are-refugees-terrorists/3489830

Africa Watch Women's Rights Project. (1993). *Seeking refugee finding terror: The widespread rape of Somali women refugees in north eastern Kenya*. Human Rights Watch.

Allen S. (1998). Identity: Feminist perspectives on 'race', ethnicity and nationality. In N. Charles & H. Hintjens (Eds.), *Gender, ethnicity and political ideologies*. Routledge.

Apple, B. (1998). *School for rape: The Burmese military and sexual violence*. Earthrights International.

Asia Pacific Women in Law and Development (APWLD). (2001). *APWLD lobby document: Women's human rights: Engendering the agenda of WCAR*. APWLD.

Bartolomei, L. A., Pittaway, E., & Pittaway, E. E. (2003). Who am I? Identity and citizenship in Kakuma Refugee Camp in northern Kenya. *Development. Xenophobia, Culture and Identity, 46*, 87–93.

Collins, P. H. (1989). The social construction of black feminist thought. *Signs, 14*(4), 745–773.

Collins, P. H. (1991). *Black feminist thought: Knowledge, consciousness, and the politics of empowerment* (1st ed.). Unwin Hyman.

Collins, P. H. (2000). *Black feminist thought: Knowledge, consciousness, and the politics of empowerment* (2nd ed.). Routledge.

Copelon, R. (1994). Surfacing gender: Re-engraving crimes against women in humanitarian law. *Hastings Women's Law Journal, 5*(2), 243–266.

Crenshaw, K. (1989). Demarginalizing the Intersection of race and sex: A black feminist critique of antidiscrimination doctrine feminist theory and antiracist politics. *University of Chicago Legal Forum, 140*, 139–167.

Crisp, J. (2000). A state of insecurity: The political economy of violence in Kenya's refugee camps. *African Affairs, 99*, 601–632.

Department of Immigration, Multicultural Affairs (DIMA). (1996). *Guidelines on gender issues for decision-makers*. Australian Government Publishing Service.

Fraser, A. S. (1999). Becoming human: The origins and development of women's human rights. *Human Rights Quarterly, 21*(4), 853–906.

Harrell-Bond, B. (2002). Can humanitarian work with refugees be humane? *Human Rights Quarterly, 24*, 51–85.

Hercus, N., Ray, N., & Pittaway, E. (2000). *The women at risk program: An evaluation proposal*. UNSW Centre for Refugee Research.

Hooks, B. (1984). Black women: Shaping feminist theory in feminist theory from the margin to the centre. In K. K. Bhavnani (Ed.), *Feminism and race*. Oxford University Press.

Hyndman, J. (2000). *Managing displacement: Refugees and the politics of humanitarianism*. University of Minnesota Press.

Kelley, N. (1989). *Working with refugee women: A practical guide*. United Nations High Commissioner for Refugees.

Mohanty, C. (1991). Under Western eyes: feminist scholarship and colonial discourses. In C. Mohanty, A. Russo & L. Lourdes (Eds.), *Third world women and the politics of feminism*. Indiana University Press.

Naik, A. (2003).West Africa scandal points to need for humanitarian watchdog. *ReliefWeb International*. https://reliefweb.int/report/guinea/west-africa-scandal-points-need-humanitarian-watchdog-0

Phillips, J., & Spinks, H. (2013, March 20). *Immigration detention in Australia*. Social Policy Section, Parliament of Australia. https://www.aph.gov.au/About_Parliament/Parliamentary_Departments/Parliamentary_Library/pubs/BN/2012-2013/Detention#_Toc351535442

Pittaway, E., & Bartolomei, L. (2001). Refugees race and gender: The multiple discrimination against refugee women. *Refuge: Canada's Journal on Refugees, 9*(6), 21–32. https://doi.org/10.25071/1920-7336.21236

Pittaway, E., & Bartolomei, L. (2002a). *Fieldtrip report to Kenya*. Centre for Refugee Research, Sydney. Retrieved September 5, 2009, from http://www.crr.unsw.edu.au/documents/Kakuma%20reportShort.pdf

Pittaway, E., & Bartolomei, L. (2002b). *Refugee women*. In Searching for Solutions Background Papers for the Executive Committee of the UNHCR Programme Annual Meeting 2002b. Centre for Refugee Research, Sydney. Retrieved June 10, 2009, from http://www.crr.unsw.edu.au/documents/Excom%20Papers.pdf

Pittaway, E., & Bartolomei, L. (2003). Where death is the only medicine—health risks. *Health Exchange, 1*, 10–13.

Real, M. J., Aggarwal, A., & Pasimio, J. (2002). *Intersectionality of women's rights*. APWLD, Chiang Mai.

Shan Women's Action Network. (1999, September). Shan women's organisation: Newsletter. Retrieved September 22, 2008, from http://www.shanwomen.org/Newsletter1.doc

Shan Women's Action Network. (2001, July). *Shan women's organisation: Newsletter*. Retrieved September 22, 2008, from http://www.shanwomen.org/Newsletter2.doc

The Guardian. (2005, February 19). UNHCR chief accused of harassment. https://www.theguardian.com/world/2005/feb/18/sarahleft

United Nations High Commissioner for Refugees. (1998). *Report of the Twelfth Meeting of the Standing Committee, 23–25 June 1998*. A/AC.96/905. United Nations High Commissioner for Refugees, Geneva. Retrieved July 14, 2009, from http://www.unhcr.org/3ae68d930.html

United Nations High Commissioner for Refugees. (2001). *UNHCR global report 2001—Kenya, United Nations High Commissioner for Refugees*. Geneva. Retrieved October 20, 2009, from http://www.unhcr.org/3dafdcf53.html

United Nations High Commissioner for Refugees. (2006, February 28). Truth overboard—The story that won't go away. *Sydney Morning Herald*. https://www.smh.com.au/national/truth-overboard-the-story-that-wont-go-away-20060228-gdn224.html?page=fullpage#contentSwap1

CHAPTER 8

Refugee Women at Risk and Reciprocal Research: 'Stop Stealing Our Stories': Beyond Do No Harm—2003–2004

Narrated by Linda Bartolomei

Differences in the Refugee women's Groups in Kenya and Thailand

The new research project was very exciting, and it felt as if at last we had an opportunity to achieve a more concrete outcome, or 'research impact', working with the refugee women and networks we had established in Kakuma and on the Thai–Burma border. The camps in the two countries had evolved very differently, and the political reasons for this had directly influenced the situations for refugee women and girls in each place. In Thailand, with the permission of the Thai authorities, the camps had been almost entirely self-managed by the ethnic Karen and Karenni refugee communities, with UNHCR having no formal role prior to 1998. They were not allowed to be designated as 'refugee camps', because this would have implied acceptance by the Royal Thai Government that the Burmese Junta was persecuting its citizens. Instead, they were designated 'camps for people fleeing fighting'. By 2003, while UNHCR and NGOs had significant roles in service provision, the refugee committee continued to play a major role in managing the camps. This enabled the refugees to retain a degree of political autonomy and, this was very male-dominated. However, Women's League of Burma (WLB) and the Karen Women's Organisation (KWO), headed by a small group of very powerful women, were playing a growing role in supporting young women's leadership.

E. Pittaway and L. A. Bartolomei,
Only Rape! Human Rights and Gender Equality for Refugee Women,
Sustainable Development Goals Series,
https://doi.org/10.1007/978-981-19-0916-0_8

This was strongly sustained by donors and INGOs, who saw it as a way of encouraging the struggle for democracy, and an indirect way of countering the behaviour of the Burmese Junta, which controlled significant oil and gas reserves. The work of WLB and KWO was critical in expanding leadership opportunities for diverse ethnic groups of refugee women from Burma, as well as spotlighting the endemic sexual violence they faced.

The north of Kenya, where Kakuma camp was located had much less political significance internationally than Burma. As a result, UNHCR and the INGOs which managed the camps and provided services were all seriously underresourced. There were refugee camp committees, often appointed by UNHCR, and these were male-dominated. In direct contrast with Thailand, there was little external support for refugee women's leadership. In Kakuma, the Jesuit Refugee Service (JRS) provided training and support for a small group of refugee women counsellors, but they were not recognised as community leaders. In Thailand much of our learning came through direct partnerships with organised and increasingly empowered refugee women leaders, in Kenya, our learning came from the many courageous women who shared their individual, harrowing stories. Their honesty directly contributed to not only further shattering the silences, but also directly challenging the pervasive myth that '*women don't talk about rape*'.

The Global Backdrop

During this period, the global political context became more unstable, conflicts continued across Africa and what would be a protracted eight-year war in Iraq had begun. Official refugee numbers had slightly decreased from the previous year. Of the 9.5 million refugees displaced, the majority were in Africa, with Liberia, the DRC, Sudan and Somalia producing the largest numbers. In 2004, escalating conflicts in Burundi and Afghanistan saw an increase in the number of refugees. Additionally, the numbers of internally displaced people continued to grow, with official reports of 5.4 million by 2004 (UNHCR, 2004). During the ongoing civil war in Sudan, ethnic African communities in the western Darfur region became the targets of genocidal violence. Villages were targeted by government-backed militia, and thousands were killed, raped, or forced to flee. The mass rape of women and girls was a core part of the genocidal attacks, with reports that, following the rape, the militia branded the women with hot irons to permanently mark them and further ostracise them from their communities (USA, 2004).

As the humanitarian community struggled to respond to the ongoing crises across Africa, a new refugee crisis emerged in the Middle East in the wake of the Iraq war. The invasion of Iraq by USA and allied forces in March 2003 would lead to the displacement of millions of Iraqi people (Saleh, 2020). Many became refugees in neighbouring countries, with the majority hosted by Syria and Jordan. At the same time, an increasing number of natural disasters added to the growing demands for humanitarian assistance. In an already deeply troubled region, an earthquake in South Eastern Iran injured, displaced and killed thousands of people (BBC, 2003). The following December, eight countries in Asia, including Sri Lanka, Indonesia and India, were struck by a tsunami which killed almost a quarter of a million people and displaced over 1.5 million. The disaster had deeply gendered consequences with as many as four times more women than men dying; because their cumbersome clothing hampered them in the water; because girls hadn't been taught to swim; and because women tried to save children or elderly family members (BBC, 2005). There was a devastating increase in incidents of rape and domestic violence in the days and months following (Rees et al., 2005). Rape, sexual and gender-based violence, in post-disaster settings was added to our list of concerns.

Eating Humble Pie! Rights and Reciprocity in Action Research

We were very surprised in early 2003, when writing to women from the Women's League of Burma (WLB) to share the good news of our research funding, we received a very lukewarm response. They agreed to meet on our forthcoming visit to Thailand but indicated that they were not in a position to commit to being part of the planned research. We had assumed that they would be as pleased as we were that the project had been funded. It soon became clear that they were both angry and disappointed that we had not even consulted them about the research and just expected they would be part of it. They said they had thought of us as friends and advocacy partners, not as academics who were coming to '*research them*'. As they shared one example after another of feeling exploited and undervalued by academic researchers, we began to feel increasingly ashamed by the arrogance of our assumptions.

> *We are really fed up with people just coming and stealing our stories and taking our photos - and we never get anything back, not even a copy of the report. Nothing ever changes.* (Refugee woman, Thailand 2003)

Many of the women in their communities were fearful of researchers and suffering from research fatigue. They were tired of being the '*objects*' of research, the results of which were never checked or shared with them (Won, 1999). Sometimes, they said '*we just lie to researchers about our situation so they will go away and leave us alone*'! They shared a recent experience in which they had assisted a researcher to interview several refugee women who had been raped by the military in Burma. They had done so on the understanding that they would jointly publish an advocacy report. However, to their deep distress and anger, the researcher had proceeded with the publication without their approval. They said it left them feeling that their '*stories had been stolen*'. They recounted multiple experiences of researchers who had not sent back promised photos or reports, or had used materials and photos without permission, putting some women in direct danger (Pittaway et al., 2010).

> *I gave a long interview and never heard another thing until a year later, someone brought me an international magazine, and there was my story and my photo with my name. It put me in even more danger.* (Karen refugee woman, Thailand 2003 in Pittaway et al., 2010)

By not consulting them or seeking their input into the project design, we had demonstrated the worst kind of white privileged academic arrogance. The proposal had been put together in a rush to meet the deadline and it had not even occurred to us to invite them to develop the proposal with us. We had simply assumed that the project would address their concerns and that they would be eager to be part of it.

Thankfully, our pre-existing relationship, and Eileen's long history of providing advocacy training to women's groups from Burma as part of the Beijing process, made it possible to salvage this critical relationship. The women were committed to addressing rape and sexual and gender-based violence in both Burma and Thailand, and therefore recognised the potential offered by a research partnership. They agreed to support the research on the basis that it included some direct reciprocal benefit for the women in their communities. They also negotiated a strict ethical agreement to ensure joint ownership of any data collected. They asked us to provide human rights and advocacy training to their members to assist them to prepare for the Beijing Plus Ten (B + 10) meeting to be held in New York in March 2005 and we willingly agreed.

Research Ethics in Action

Our next stop was the Karenni camps near the small town of Mae Hong Son on the Thai–Burma border to meet with refugee leaders and the Jesuit Refugee Service (JRS), one of our formal NGO research partners. Having reflected on the concerns shared by WLB, we wanted to ensure that the planned research in the camps included some direct benefits for the community. This idea was well received by the JRS team, who shared similar concerns about refugee communities being over-researched, and overstretched local organisations feeling exploited by the demands of facilitating research projects. JRS provided educational services in the camps, working with the Karenni Refugee Education Committee (KNRC) and its network of refugee teachers. It was decided that we could contribute to the KNRC training program, by providing two Training of Trainer (ToT) Programs, one on gender and one on human rights, for some of the refugee teachers. They advised that an experiential learning approach would be most useful and requested materials the teachers could use in the camps. They also agreed that we could seek the consent of the participants to document the training to inform our research.

Back in Australia—Dance Parties and Training Materials

These experiences catalysed some deeper critical thinking about our emancipatory and participatory action research principles. But, in the short term, two new training courses needed to be developed and delivered, with no funding! Out of the blue, one of our students[i] offered to fundraise for the project by hosting a dance party. The party was a huge success and raised almost $15,000, and so the work began.

To address the ethical concerns raised by the women, a confidentiality agreement was drafted between us and the groups with whom we were working, stating that we would not publish any of the individual stories shared by the women until their organisations had published their own reports or had given us written permission to do so. There was a commitment to share copies of all reports for approval by nominated organisational representatives prior to publication. This was risky as the right to publish could potentially be denied, directly threatening the expectations of our academic work. However, ethically, it was the only course of action with which we were comfortable. In fact, rather than constraining our academic work, it had the opposite effect. It led to deep and long-standing relationships of trust and a far greater understanding of the issues than might otherwise have been achieved.

Carole Shaw, an ex-student and part of the UNSW CRR team, joined us to assist in creating and delivering the training. She worked with a small group of students to develop materials for the gender and human rights training. They contributed creative ideas to teach human rights and the core principles of universality, indivisibility and inalienability. These included using multicoloured beads, with different colours for different human rights, to make a symbolic human rights necklace. Carole proposed the use of the storyboarding technique, commonly used in film and strategic planning, as one of the research methods. It involves participants working in small groups to identify issues of concern and then drawing a series of pictures in response to six questions exploring particular problems, impacts on individuals, families and communities, an assessment of current responses, and suggestions for change. Each group then presents their storyboards, sharing their analysis of the problem, recommendations and potential solutions. These are fully documented and form a core part of the research data (Pittaway & Bartolomei, 2009) (Fig. 8.1).

Developing Reciprocal Research: Beyond Do no Harm

Returning to Mae Hong Son in November, we initially agreed to conduct an intensive four-day training for 20 refugee teachers. This was quickly extended to 40, and on the first day of training, 122 people arrived. Some had walked for hours over the mountains from a remote camp. No one was turned away, and because they so desperately wanted to learn, it worked! Strict ethical agreements were negotiated and all participants, including the interpreters, signed a group confidentiality agreement. This created an environment of trust which encouraged the participants to frankly share their issues of concern. These included stories of extreme

Fig. 8.1 Example of a Storyboard: SGBV risks and impact on women and girls (India 2008)

hardship, of gender-based violence and the frequent kidnapping of children by the militia groups. The storyboard technique proved to be hugely successful as it provided participants the opportunity to engage in deep analysis of the problems and to take the lead in proposing workable solutions.

Drafts of the report were shared with nominated representatives of each group for approval, prior to publication. It was emailed in the form of a PowerPoint presentation to JRS in Thailand. A nun who was heading the local office then travelled into the camp with her laptop to share the draft presentation with nominated representatives of KNRC and other organisations which had participated. This letter from KNRC gave us permission to circulate the report (Fig 8.2). The TOT Program about gender and human rights was very well received and was used extensively in the camp schools and with peer support groups.

KnED Karenni Education Department

"Education For the Progress of People"

Date November 3, 2004

Dear Eileen Pittaway and Linda Bartolomei,

On be half of Karenni teachers and students, I would like to thank and give the respect to you. Because of the Human rights and Gender Training facilitated by you in last year, teachers and post high school students positively enriched their knowledge and shared to community.

Perfectly, we have seen the context and information about the training on CD PowerPoint for Karenni teachers and students and we are very enthusiastic for that you may use it for training (or) workshop in any areas you want. Wonderfully, we hope that international refugee community or other will see how we keenly involved in training as a part of international activities in Human Rights and Gender issues.

Honestly, we would like to keep this research document as a reference or document in KnED and other training centers. Please if it is possible, we would like you to arrange 3 copies for us.

Thank for your solidarity

(Augustino)

Director of KnED

Fig. 8.2 Letter of consent from KNRC

Building Trust: Being Heard

This new approach reflected our commitment to engage in an ethical, respectful, and reciprocal research process. It fostered a level of community trust, which enabled participants to frankly share the problems they were facing. Prior to the training, our partners had heard 'whispers' of problems, but there was a reluctance to speak openly about things such as domestic violence, trafficking and child soldiers. However, in this training, given a choice to identify priority issues of concern, several groups elected to prepare storyboards on these issues, naming them as human rights abuses and analysing both the causes and impacts (Fig 8.3). These were powerful research outcomes. This methodology was exactly what was needed to conduct research with marginalised and disadvantaged groups who had valid reasons for distrusting people such as researchers, academics, representatives of government and other institutions.

Following this successful trial, materials were adapted and extended, as a core part of the methodology called *Reciprocal Research* (Pittaway

Fig. 8.3 Storyboard—Risks for children, Thai–Burma border, 2003

& Bartolomei, 2009). It is a participatory action research approach, grounded in human rights principles, which supports research that is empowering, that is not harmful or exploitative, and that has the potential to bring about social change (Reason & Bradbury, 2006). The reciprocal nature of the method transforms people from subjects to participants in research. It moves from the notion of 'harm minimisation' as the ethical base of research to the notion of reciprocal benefit, and community-directed outcomes. The methodology had much success in both eliciting information and giving something back to community (Pittaway & Bartolomei, 2009; Pittaway, et al., 2010).

Creating Safe and Respectful Research Spaces

As the training progressed, we endeavoured to foster a sense of safety and protect privacy as best we could. Many of the camp-based consultations were held in flimsy bamboo structures, close to the huts in which many refugees lived. There was no privacy, so if we received warning of government 'spies' being in the camps, microphones were turned off and the workshop was conducted in low voices—or we began singing. On occasions, posters were quickly removed, and the topic changed from human rights to something considered less subversive, such as maternal health care. Nominated local staff provided individual debriefing and counselling to participants if required, and activities affirming community strengths were included each day. There was always a final celebration, sometimes light-hearted, and at other times a ceremony acknowledging their resilience and strength. As a gesture of respect and recognition, each participant received a certificate of participation. Often this simple act brought participants to tears, as it was not only the first certificate they had ever received, but the first document. Many refugees have lived their whole lives without even a birth certificate.

Reciprocity and Shattering Silences

After Mae Hong Son, we provided the five-day human rights and advocacy training to the members of WLB in Chiangmai. This was held in a large, soulless, but very busy, three-star business hotel. It was specifically selected to provide the refugee women with a safe space where they could blend in with the crowds of guests. As it was the first time many of the women representing WLB's 13 member organisations had met us, we spent the first day working to establish our credibility as trainers and researchers. Late that evening, there was a knock on the door and one of the very well-respected senior Karen Women's Organisation (KWO) leaders asked if she could speak to us privately. Apologising for the lateness, she came straight to the point. She handed us a box of computer disks and asked for help to prepare an advocacy report. The disks contained first-hand accounts from ethnic Karen women and girls who had been raped by the military in Burma. They had been collected over many months on dangerous, clandestine trips across the border. Deeply moved by this gesture of trust, we felt a strong moral obligation to assist. This was the first test of whether our team would live up to its promise of reciprocity.

Drawing on our resources,[ii] we worked with KWO to draft the report *Shattering Silences* (KWO, 2004). It was published in April 2004 and the women used it as an advocacy tool to raise international awareness of the systemised rape of women and girls in conflict. Most significantly, it was directly quoted by the then US Secretary of State, Condoleezza Rice, in 2008, to argue that rape *was* an international security concern and must be included on the Agenda of the UN Security Council. UNSC resolution 1820, directly addressing this, was adopted in the following meeting (UNSC, 2008). We next flew to Kenya!

Kakuma Camp, Kenya: A Powerful Participatory and Emancipatory Research Process

> *This is a rape camp. It does not matter what you wear. If you wear shorts, they will rape you. If you wear Muslim dress, they will rape you. This is a rape camp.* (Teenage refugee girl, Kakuma, 2003 in Bartolomei, 2010)

Contrary to the common wisdom that refugee women were unwilling to disclose incidents of rape and sexual violence (UNHCR, 1995), women in Kakuma queued for hours to meet with us to share their stories. One of the keys was that we had established trust with the women and provided a safe space for them to do this. *Reciprocal Research,* with the interactive introduction to human rights provided the safety, opportunity, and permission for groups of refugee women and men to share their individual and collective experiences of abuses. The most common response we heard was: '*This is the first time I have been allowed to tell my story – this is the first time anyone has listened to me*'.

Over the next two years, we conducted hundreds of in-depth interviews with women in both Kakuma and Nairobi. Most had been raped not only once, but on multiple occasions, in their home countries, during flight and again in Kenya. Some had been held as sex slaves, others reported gang rape, and many had given birth to babies of rape (Bartolomei, 2010; Pittaway et al., 2003). Far from being a place of safety and asylum, women and girls remained at heightened risk in the camps. However, what was emerging was that the risks for some women and girls were far greater than for others.

Women and Girls at Heightened Risk

Unaccompanied minor girls informally adopted by foster families were one of the most at-risk groups identified. Contrary to their own guidelines, UNHCR appeared to have no capacity to play a significant role in, or supervision of, these foster arrangements (UNHCR, 1995, p. 2.1i).

> *My foster mother had many sons and they started to abuse and rape me. I complained to the woman, and she got very angry with me and defended the boys. She said you are born of an unmarried mother and you are a prostitute in my house. She threw me out of the house, and would not give me any wages, or my share of food rations. I was on my own, aged 14 years.* (Sudanese refugee woman, Kakuma, 2003 in Bartolomei, 2010)

The JRS staff had identified that those women suffering from severe trauma because of previous experiences of rape were particularly vulnerable to further rape and sexual abuse. Social exclusion, stigma and fear of further attacks had forced many women to flee to the 'Safe Haven' (Safehouse).

> **How the System Failed Women at Risk**
> *Miriam*[1] *had been raped twice in Kakuma and was suffering from severe trauma and depression. She had been referred by UNHCR to Nairobi for psychiatric assessment and on return to the camp had been referred for assessment as a woman at risk. Shortly after her return to the camp, she was raped*

a third time and became pregnant as a result of the rape. As a consequence of her pregnancy she was rejected by her family and community. The fact that Miriam had already been raped twice in the camp and was only then being assessed as a woman at risk points to the inadequacy of the protection response. That she was sent all the way to Nairobi for a psychiatric assessment only to return to be raped a third time, highlights a chronic failure of any form of protection system. She was clearly already deeply psychologically scarred and at heightened risk as a result of the two previous rapes but now, having been raped a third time, she was denied what little social support she might have enjoyed by being able to remain with her community and family. Without this support the degree of risk she faced in the camp increased. Stigma, shame, and social exclusion intersected with other key risk factors to increase her vulnerability to further incidents of abuse.

Despite the UNHCR Guidelines (1995), there were no specific programs or policies designed to identify those potentially at risk when they arrived in the camps. UNHCR staff struggled in the face of overwhelming need and had complained that one of their biggest difficulties was that they lacked the systems to prioritise those in greatest need of protection. In fact, despite the rhetoric of referral systems and response pathways shared by several service providers, there appeared to be no functioning systems at all.

Protection for Women at Risk? The Sacred and the Profane

One of the biggest challenges to address was how service providers downplayed the extent of the problem and overplayed the extent and quality of the responses available. For example, a UNHCR staff member we met on our second visit challenged our focus on refugee women at risk when there were so many other important priorities in the camp. However, meeting her one evening in the informal bar in the UNHCR compound, she told a very different story. Defences down, she admitted she simply could not cope with acknowledging or talking about the appalling levels of rape and sexual violence that women and girls faced every day in the camp. Similarly, the one service provider responsible for providing services

[1] Not her real name.

to the survivors of gender-related violence in the camp of 95,000 people, claimed there were clear referral pathways and case management systems in place. Asked how often the case management meetings occurred, she paused and said, 'well they are supposed to be once month, but… ah… we did not have one last month because the UNHCR gender focal point was on leave, and we won't have one this month because I will be taking leave'.

It quickly became clear that UNHCR's bureaucratic structure was problematic: Protection, Community Services and Resettlement staff all worked in 'silos', with different reporting lines and responsibilities and little communication between the departments. With the permission of several women then sheltering in the Safe Haven, who clearly met the criteria for a WaR Visa, we asked UNHCR to review their applications for urgent resettlement. The officer in charge of resettlement looked dismayed, as she had not heard of any of their cases. Immediately on the defensive, the protection officer told us that there were so many women who fit the criteria that it was impossible to prioritise referrals. This was compounded by the lack of a shared filing system within the office. By working in silos, the resettlement and protection staff were unaware of cases in need of either urgent protection assistance or resettlement. The systemic failures were clear.

NGO Camp Culture and Attitudes

Another deeply disturbing phenomenon noted among NGOs was the propensity of many NGO workers to unthinkingly propagate myths and anecdotes about the refugee populations as fact, leading to a confusing pastiche of truth and misinformation. This included comments such as '*the Sudanese come to the camp deliberately to get education and training*', the implication being that the Sudanese were not 'real' refugees. This narrative denied the reality from which the Sudanese people had fled, and the danger they faced in returning. In other cases, the extremist Christian views of some staff dictated their service provision. Staff of one large INGO refused shelters to unmarried young women with babies because 'they were immoral'. There was also implicit pressure on refugee recipients to convert: a staff recruitment notice, from the INGO, read '*local staff required to work as drivers, must be born again Christian*'. The indomitable Sr Christina, the Mercy nun who ran the JRS Safe Haven, informed us that internal conflict between different extremist Christian groups was a major problem in the camp, and that recruitment practices such as these only fuelled the conflict, to the detriment of refugee women. It was all a very long way from anything that even approximated a human rights-based approach to service provision.

The WaR Program Continues to Fail

It was clear that a vastly more efficient system of identification and WaR case processing was urgently needed. Interviews with senior UNHCR staff in Nairobi confirmed these problems.

> 'One of the greatest difficulties UNHCR faces is in identifying those women facing high levels of risk. So the processing of WaR cases generally takes longer – as the definitions are ambiguous and the cases often fall between Protection and Social Services – their roles are not clear' (UNHCR Interview 2003 in Bartolomei 2010).

UNHCR staff comments about the ambiguity of the definition of WaR echoed those made by UNHCR in Thailand. However, it was clear that this was not the only factor, as the lack of systems at camp level contributed to the difficulty in identifying and processing WaR cases. Negative staff attitudes were clearly a major problem. This was exacerbated by the time it took between referral and resettlement to a third country. In April 2002, the Australian Government had committed to resettle 700 refugees. However, in June 2003, 16 months later, many places were unfilled. This included WaR cases which, in theory, should have been fast-tracked for resettlement once identified. Many WaR in dangerous situations remained at extreme risk of further incidents while they waited for resettlement. Once it became known that they had been accepted for resettlement, women who had fled forced marriages and domestic violence also faced risk of abduction, or having their children kidnapped by their husband's families. In one case, a woman was raped and subsequently gave birth while waiting to be resettled. When she was finally given a travel date, she was told that she could not take her baby as the child was not included on her resettlement application. She was advised to depart and to sponsor her child once she arrived in her resettlement country.

Addressing the Systemic Failures of Protection

The incidence of rape and sexual violence was so widespread and entrenched that the problem seemed too big to solve. We sat discussing this late into the night, locked in our dusty compound in Kakuma camp, behind razor wire, with people guarding us with guns. Eileen had been thinking about the comment that had been made by the UNHCR Head of Camp on our previous visit: '*I will give you a prize if you can find a woman in this camp who has not been raped*'. If all of the women and girls faced unacceptably high levels of risk, what would an effective method of identifying those at extreme risk look like?

The definition of WaR in the Resettlement Handbook (2002) provided no clear guidance. It simply identified a range of problems women might face, including 'security threats, sexual harassment, violence, torture and different forms of exploitation' and noted that the loss of traditional family structures, including the lack of support of an adult male, might make some women more vulnerable than others (UNHCR, 2002 Handbook, IV p. 13). As this appeared to apply to a majority of women and girls in Kakuma camp, we began to question the usefulness of the WaR Program which, in Kakuma, might at best offer resettlement to 1% of the women in the camp (Crisp, 2000).

Action was imperative, but what action and how? A step back to get critical distance was needed. UNHCR and the NGOs could do better. There had to be a more effective method of identifying those most at risk. New tools and clearer guidelines were needed. In exasperation I asked, 'How on earth could another tool and more guidelines possibly help, given the plethora of guidelines that UNHCR staff had and didn't appear to use'? The findings from the fieldwork in both Kenya and Thailand only confirmed the range of risks for women and girls that had already been so clearly laid out in the *Guidelines on Sexual Violence against Refugees* (UNHCR, 1995) and reconfirmed in their 2001 review (UNHCR, 2001). Staff at UNHCR Nairobi had told us that each year brought more and more guidelines and fewer and fewer resources to support their implementation (Bartolomei, 2010). I doubted that they would welcome yet another.

Nonetheless, one very early morning in Kakuma, I sat at my laptop and began the first draft of an individual risk assessment tool that has now been expanded and adopted by UNHCR as standard practice.[2] It was not much more than a checklist, but as I typed I remember feeling a small glimmer of hope that a more methodical approach might be possible. The checklist developed was a two-part list; the first block of factors was designed to flag potential risk, including women's previous experiences of abuse or characteristics and circumstances which might increase their risk of abuse. The second block focused on her current circumstances and was designed to identify whether she was currently experiencing or was in immediate danger of any of the abuses listed.

WaR Risk Factors Checklist—Draft, Kakuma Camp, 15/06/03

A. Reasons for possible or likely persecution of women and girls in Kakuma camp.

- Political affiliations or those of husband/family
- Rape victim/survivors
- Mixed marriage (across ethnic or clan lines)
- Single women/unaccompanied minors
- Forced marriage
- Threat of abduction (self or children)
- Exercised choice of sexual/relationship partner

B. The second part assessed whether the woman faced these or additional risks in the camp.

Risk factors in camp.

- Rape and sexual violence
- Lack economic support/livelihood opportunities
- Lack free movement in camp – imprisonment
- Battery and other violence – leading to death
- Harassment – of children, stoning of self or house
- In state of constant fear—psychological pressure
- Rejection or persecution by own community
- FGM

[2] The UNHCR version of the Tool is the Heightened Risk Identification Tool (HRIT).

- HIV/AIDS & other STDS
- Forced Prostitution/(trafficking)
- Abduction of self or children

If no effective protection can be provided from any of the risks in B, then the woman or girl should be resettled under the WaR Program.

The development of this two-part list was a precursor to a deeper analysis of the way that risk, and vulnerability to risk, intersect and increase the susceptibility of women to further incidents of rape and sexual violence. Structuring the checklist in this way facilitated an analysis of the way previous experiences of rape and sexual abuse increased women's risk of further incidents of violence, including risks of domestic and family violence and of being forced to engage in survival sex. This included the role which the shaming, stigmatisation and social exclusion of women and girls who have been raped, played in increasing women's risk of further incidents of abuse.

From the Horrors of Kakuma to the Affluence of Geneva

From Kenya, we travelled directly to UNHCR Geneva to attend the UNHCR June 2003 Standing Committee meeting, aiming to share our preliminary research findings and advocate for action. To do this we needed to learn much more about the UNHCR policy-making process and how, as academic activists, we could influence it.

This was my first visit to Geneva. I was excited but also anxious. Out came the 'UN-appropriate wardrobe', which turned out to be way too warm given the early arrival of summer that year. So, it was back to the casual, cotton 'Kakuma' wardrobe. Contrary to my expectations, the lack of experience with formal UNHCR processes and utterly inappropriate clothing proved not to be a weakness, but added to our credibility, as we had '*actually been in a refugee camp*'. After the second day, I was amazed at the number of NGO and UNHCR staff who commented on this. Naively, I had assumed staff working in international policy would have direct experience of, and regular contact with, the field. However, the opposite was true. Many had moved directly from junior policy-making roles in their home countries to positions of significant power within UNHCR or to head powerful INGOs with little actual experience in the field. These were the 'policy people', priding themselves on knowing the rules of engagement in UNHCR's international policy processes. It was a cosy 'club' of a dozen or so mainly white representatives of large INGOs which were UNHCR's major implementing partners. Like any 'insider' club, it had developed a culture and a series of mainly tacit rules of membership. The most significant was that UNHCR was a humanitarian agency and the meetings of the UNHCR Executive Committee (ExCom), while providing a formal space for invited NGOs as observers, were not, unlike the Human Rights Council, a political space. Eileen had, however, already begun to shake things up the year before.

Politicising the Advocacy Space at UNHCR Geneva and a Game of 'Yes Minister'

Drawing on experience in the Commission on the Status of Women, in 2002, a group of Australian NGOs, refugee rights activists and students, led by Eileen, participated in the UNHCR Pre-ExCom and ExCom meetings as part of the recently formed Australian Refugee Rights Alliance[iii] (ARRA). ARRA formed following the *Refugees: Where to from Here?* conference convened at UNSW in 2001. Its aim was to exert pressure on the Australian Government to end the mandatory detention and offshore processing of refugees. As efforts in Australia were slow to find traction, and believing in a multi-pronged approach to advocacy, we encouraged the ARRA group to focus their efforts internationally. Eileen called for volunteers to self-fund to attend the meeting, as, with her ECOSOC accreditation, she could sign them in. Twelve people volunteered to go.

ARRA's arrival on the UNHCR Geneva scene really caused a stir. UNHCR's NGO liaison unit, while supportive of the team's participation, was nervous about whether such a large group of newcomers would play by the agreed rules. UNHCR's Executive Committee (ExCom) is composed of member states, both signatories and non-signatories to the 1951 Refugee Convention, whose role is to provide UNHCR advice on its programs and activities. As UNHCR receives significant funding from many of these states, it is forced to walk a fine line between criticising governments failing in their international obligations to refugees, and encouraging them to do more. Getting the balance wrong can lead to governments cutting or withdrawing their annual contributions. The last thing UNHCR wanted was a mob of unruly NGOs disturbing the precarious rules of international diplomacy! Their anxiety only increased when a representative of the Australian Government contacted UNHCR for a list of their names. This was a slightly unusual request (from a democratic nation), and one which raised concerns within UNHCR that their approval processes for NGO registration might be questioned. Eileen calmed UNHCR anxieties, assuring them that she had correctly registered all participants as members or affiliates of organisations who had ECOSOC status, meaning they were formally accredited to participate in UN meetings. In fact, it would have been difficult for UNHCR to prevent them attending. She also contacted the Australian mission to enquire what details they required and why. A somewhat ruffled assistant said he would need to check. He called back a few days later, advising that the then Immigration Minister, Phillip Ruddock, was attending the meeting and was keen to meet with the Australian NGO delegation.

Invitations to a cocktail party soon followed. This left the ARRA team in a quandary; many simply did not wish to accept the invitation. The group came up with a clever strategy. They agreed that it was most likely that the Minister was expecting them to challenge him about Australia's asylum policy, so they agreed they would not mention it. Instead, they would question him on Australia's contingency plan for the refugees who would be generated should the then-planned attacks on Iraq proceed. When they met with the Minster at the cocktail party he was harried and irritated when they did not raise asylum policy, for which he obviously had a prepared statement. '*So why did you want to meet me*'? he asked. '*But Minister*', replied Eileen '*we did not ask to meet with you, it was you who requested the meeting with us*'! Somewhat taken aback, he stumbled on. However, the group's question about potential refugee flows from Iraq left him speechless. It was clear that he had not even considered it and equally clear that he had no real desire to meet the group. It would appear their names might only have been requested in an attempt to block their participation at the UNHCR ExCom meeting.

Building Our Credibility on the Geneva Scene

The ARRA team quickly gained a reputation for being well organised, credible and politically savvy advocates. It was clear from our reception on arrival at the UNHCR Standing Committee in June 2003 that we had been accorded a level of 'discursive legitimacy' by some of the more experienced Geneva-based NGOs and senior UNHCR staff. It is, however, important to note that the team who participated in the

UNHCR meetings in the first two years did not include anyone from a refugee background. This was an obvious gap and, from 2004, with a network of supporters willing to donate funds for their airfares and accommodation,[iv] an increasing number of women and men from refugee backgrounds joined the team each year.[v]

ExCom Conclusions on International Protection

One of the roles of UNHCR's ExCom is to identify new and emerging issues of concern relating to the protection of refugees and to support the development of new policy and practice frameworks. While a number of gaps and problems might be discussed within the context of the general meeting debate, the main means through which states and UNHCR seek to address these gaps is through the issuing of Conclusions on International Protection. These are designed to address gaps within UNHCR's existing law and policy framework (Sztucki, 1998). They aim to expand the protection available to refugees and asylum seekers through addressing key omissions in the 1951 Convention, and by ensuring that the legal framework underpinning refugee protection effectively addresses contemporary issues. Although they are so-called soft law and are therefore not in themselves legally binding, their strength lies in the fact that they are consensus documents which have been signed by the Member States of the Executive Committee.

A 'Lightbulb Moment' at the UNHCR Protection Standing Committee Meeting

There are three preparatory 'Standing Committee' meetings held in March, June, and September prior to the October ExCom. As with all major UNHCR documents, the text of the Conclusions is debated and agreed on well in advance of each ExCom meeting. At the June Standing Committee, the Conclusions to be adopted that year were still being debated, and we were pleased to see that the draft text for the General Conclusion included a strong reference to the importance of applying a gender lens in all UNHCR's policy-making. We recognised this as an important advocacy opportunity and vowed to learn more about how we might seek to influence this process. We discussed whether in future we should push for a Conclusion on Women at Risk, doubting that it would actually be possible, as we were told that civil society had no role in drafting Conclusions. Undeterred, we held a number of informal meetings with senior staff within UNHCR, who became strong allies about the possibility of a Conclusion on Women and Girls at Risk the following year. The advocacy efforts of the ARRA team had been well received by Erika Feller, then Director of UNHCR's Division of International Protection (DIP). We met with her to provide an update on the progress of the WaR research, and she promised ongoing assistance. We also had an important dinner with the then Senior Coordinator on Women, Joyce Mendes-Cole, who recognised us as gender allies and offered to assist us.

Back in Geneva for the ExCom Meeting in October 2003

Having secured one of the limited timeslots available for a side meeting during the Pre-ExCom NGO Consultations in September 2003, we invited Erika Feller to launch *Seeking Hope, Seeking Safety,* a booklet of case studies of women at risk highlighting some of the major problems with the WaR Program (CRR/ANCORW, 2003). *Welcome to Kakuma*, the twenty-minute film we had produced a year earlier to highlight the negative impacts of prolonged encampment, in particular on women at risk, was also screened. Erika Feller commended not only the preliminary findings of the research, but also the effectiveness of the film as an advocacy tool. She requested a copy for UNHCR to use in their own discussions with governments to mobilise additional resources and improved responses for refugees in protracted refugee settings.

Moving Forward: Lessons Learned in This Research and Advocacy Cycle

By using the Reciprocal Research approach we privileged the knowledge and expertise of the refugee women's groups, emphasising the mutuality of the research process (Dominelli, 2002; Ife, 1997). We did this by responding directly to the priorities and concerns of the refugee women's groups rather than rigidly following our own research agenda. In fact, there were many dramatic diversions from the original research plan. The initial narrow focus on the WaR Program was expanded to engage with the broader protection risks faced by women in both Thailand, Burma and later in Kenya. This provided the opportunity to share technical knowledge, and experience of advocacy within the UN system, with the refugee women's organisations to empower them to advocate for themselves. The reciprocity in our research approach is informed by anti-oppressive social work principles. This approach is underpinned by values of reciprocity, equality and democracy and works holistically to promote social justice and individual and collective empowerment (Dominelli, 2002). It involves a constant revaluation of one's own power and perspectives, requires knowledge and skills to challenge racism, nurture emancipation and to work across cultural and linguistic difference (Dominelli, 2002; Quinn, 2003). These approaches to individual and community empowerment are underpinned by understandings that this is not something 'done to' individuals or communities but that instead to work with communities to create the environmental and social conditions in which community capacities that enable self-determination can grow and flourish (Kenny, 2006; Westoby, 2008). This is not a straightforward process and at times we struggled to stay true to the philosophical base, but the transparency inherent in the tools which we developed assisted us to stay honest!

The Power of Human Rights Trainings

At times, INGO staff questioned the merit of providing human rights training in refugee communities where so many factors prevented them from realising either their civil or political rights. However, others could see the merit of this training.

> I've done a lot of training to empower women but at the same time to educate men. Women's groups are trying to adopt laws against domestic violence in the camps. Things are changing these last three years; the attitude of the camp leaders has gradually changed; the second step is to put good women into decision-making positions to strengthen the changes... Most important is the confidence it [human rights training] gives – this makes a huge impact. It empowers them to take on their own problems. (Burmese Human Rights Trainer, Thailand, 2002 in Bartolomei, 2010)

A rights-based approach underpinned the trainings and community consultations conducted with members of the refugee communities in Thailand during 2003, and in subsequent years in Kenya. The aim was to engage with community members as respected partners, each of us making important and reciprocal contributions to the training process (Escobar, 2001).

Trainers were able to share knowledge of the human rights framework, and community members contributed their knowledge of the manner in which it might be adapted and applied to the problems they were facing in their communities. Through this process of respectful exchange, both women and many men in the communities displayed their clear preparedness not only to acknowledge the high incidences of rape and sexual violence, but also to address it. Principles of autonomy and self-determination are central to the human rights framework (ICCPR & ICESCR, 1966a, 1966b, Article 1) and within empowerment-based approaches to community development practice. 'Change from the grass-roots' highlights the critical importance of centring and valuing the skills, knowledges, capacities and experiences of communities (Ife, 2002). Participation was recognised as a right and is integral to the process of community development.

A number of humanitarian aid actors critiqued community development's commitment to 'participation', arguing that it was more rhetoric than reality (Cornwall & Brock, 2005; Slim, 2002). In the development and implementation of a genuine rights-based practice, however, an emphasis on participation is essential. Thus the language of a Participatory Rights-Based Approach (PRBA) was appropriate to describe our approach. The point here is that if the affected communities are not actively participating in identifying rights violations and solutions, it cannot be a rights-based approach.

While both UNHCR and the INGOs paid lip service to community-based approaches, little in their practices reflected a genuine understanding of either a rights-based or community development approach (Ife, 2001). UNHCR's seemingly perennial challenge to achieving effective refugee participation continued. Meanwhile, the refugee community-based organisations had made significant progress in identifying and responding to problems for women in the camps and urban areas. Using *Reciprocal Research* in community consultations provided genuine opportunities for the active and ongoing participation of marginalised and excluded refugee communities. It provided forums in which refugee communities could describe and analyse their problems, develop appropriate solutions and remain actively engaged in their ongoing implementation.

Next Steps

It was both an exciting and a deeply challenging time, as we moved from the atrocities of the refugee camps to the affluence of Geneva and back to our own homes and lives in Sydney. We had developed partnerships with NGOs, refugee women's organisations and UNHCR staff which were enriching at both a personal and professional level. A nascent respect for us as academic experts at UNHCR Geneva increased our ability to influence international policy. Most importantly, we met and worked with an incredible team of refugee women, many of whom would become lifelong friends in the continued struggle for refugee, and indeed for all, women's rights. These friendships and strategic organisational alliances grew stronger over time. As a result of our work in Kenya and Thailand, UNHCR Geneva began directly funding our research and our work expanded into other sites in Africa and Asia.

Notes

1. Our thanks to David Green and the many students who supported their efforts.
2. Alison Aggarwal, Emma Pittaway and Colleen Bartolomei provided assistance.
3. ARRA founding members included Sr Aileen Crowe, Jacquie Everitt, Jackie Keegan, Olivia Wellesley- Cole, Diana Goldrick, Kerrie Blackwell and Jessica Rossell.
4. Several non-refugee ANCORW board members provided significant financial support over a number of years to enable the refugee women on the board to participate. Paul Power from RCOA also worked tirelessly to raise funding to support refugee representation.
5. In 2005 Juliana Nkrumah from Ghana joined the team, in 2006 Tennah Kpaka from Sierra Leone participated and in every subsequent year between two and four women with lived experience were sponsored by ANCORW to participate.

References

Bartolomei, L. (2010). *Struggling against the silences: Exploring rights based responses to the rape and sexual abuse of refugee women and girls* (Unpublished doctoral dissertation). UNSW Sydney.

Bartolomei, L. A., Pittaway, E., & Pittaway, E. E., (2003). Who am I? Identity and citizenship in Kakuma Refugee Camp in northern Kenya. *Development. Xenophobia, culture and identity, 46*, 87–93.

BBC. (2003, December 26). *Iran earthquake kills thousands*. British Broadcasting Corporation. http://news.bbc.co.uk/2/hi/middle_east/3348613.stm
BBC. (2005, March 26). *Most tsunami dead female—Oxfam*. British Broadcasting Corporation. http://news.bbc.co.uk/2/hi/asia-pacific/4383573.stm
Centre for Refugee Research (CRR)/ANCORW. (2003). *Seeking Hope Seeking Safety*. ANCORW.
Cornwall, A., & Brock, K. (2005). *Beyond buzzwords: "Poverty reduction", "participation" and "empowerment" in development policy, overarching concerns*. Programme Paper Number 10, November 2005. United Nations Research Institute for Social Development. Retrieved September 10, 2008, from http://www.equalinrights.org/
Crisp, J. (2000). A state of insecurity: The political economy of violence in Kenya's refugee camps. *African Affairs, 99*, 601–632.
Dominelli, L. (2002). *Anti-oppressive social work theory and practice*. Palgrave Macmillan.
Escobar, A. (2001). Planning. In W. Sachs (Ed.), *The development dictionary* (pp. 132–145). Witwatersrand University Press.
Ife, J. (1997). *Rethinking social work: Towards critical practice*. Longman.
Ife, J. (2001). *Human rights and social work. Towards rights-based practice* (1st ed.). Cambridge University Press.
Ife, J. (2002). *Community development: Community-based alternatives in an age of globalization* (2nd ed.). Pearson Education.
Karen Women's Organisation (KWO). (2004). *Shattering the silences: Karen women speak out about the Burmese military regime's use of rape*, KWO, Thailand. https://karenwomen.org/2004/08/27/shattering-silences/#:~:text=Released%20in%202004%2C%20Shattering%20Silences,for%20Karen%20women%20in%20Burma
Kenny, S. (2006). *Developing communities for the future* (3rd ed.). Thomson.
Pittaway, E., & Bartolomei, L. (2009). Innovations in research with refugee communities. *Transitions*, 21. NSW Service for the Treatment And Rehabilitation of Torture And Trauma Survivors (STARTTS). Retrieved September 10, 2009, from http://www.startts.org.au/default.aspx?id=405
Pittaway, E., Bartolomei, L., & Hugman, R. (2010). "Stop stealing our stories": The ethics of research with vulnerable groups. *Journal of Human Rights Practice, 2*, 229–251.
Quinn, M. (2003). 'Immigrants and refugees: Towards anti-racist and culturally affirming practices'. In J. Allan, B. Pease, & L. Briskman (Eds.), *Critical social work*. Allen & Unwin.
Reason, P., & Bradbury, H. (2006). *Handbook of action research*. Sage.
Rees, S., Pittaway, E., & Bartolomei, L. A. (2005). Waves of violence—Women in post-Tsunami Sri Lanka. *Australasian Journal of Disaster and Trauma Studies, 2*, 1–6.

Saleh, Z. (2020). *The human cost of U.S. interventions in Iraq: A history from the 1960s through the post-9/11 wars*. Costs of War. Watson Institute International & Public Affairs. Brown University.

Slim, H. (2002). Not philanthropy but rights: The proper politicisation of humanitarian philosophy. *The International Journal of Human Rights, 6*(2), 1–22. https://doi.org/10.1080/714003759

Sztucki, J. (1998). The conclusions on the international protection of refugees adopted by the executive committee of the UNHCR programme. *International Journal of Refugee Law, 1*(3), 285–318.

U.S. Department of State. (2004). Sudan: Ethnic cleansing in Darfur—Fact sheet. 60th Session of the UN Commission on Human Rights. Bureau of Democracy, Human Rights and Labor. U.S. Department of State. Retrieved April 27, 2004, from https://2001-2009.state.gov/g/drl/rls/31822.htm

United Nations (1966a). *International covenant on civil and political rights (ICCPR).*

United Nations (1966b). *International covenant on economic, social and cultural rights (ICESCR).*

United Nations High Commissioner for Refugees (UNHCR). (1995). *Sexual violence against refugees: Guidelines on Prevention and Response.* United Nations High Commissioner for Refugees.

United Nations High Commissioner for Refugees (UNHCR). (2001). *Prevention and response to sexual and gender-based violence in refugee situations, interagency lessons learnt conference, 27–29 March.* United Nations High Commissioner for Refugees.

United Nations High Commissioner for Refugees (UNHCR) (2002). *Resettlement handbook* (3rd ed). Division of International Protection, United Nations High Commissioner for Refugees.

United Nations High Commissioner for Refugees (2004). *UNHCR statistical yearbook country data sheet—Thailand.* Retrieved October 15, 2009, from http://www.unhcr.org

United Nations. (2008). *Resolution 1820, adopted by the security council at its 5916th meeting, on 19 June 2008.* S/RES/1820 (2008), United Nations, New York. Retrieved November 20, 2008, from http://www.un.org/Docs/sc/unsc_resolutions08.htm

Westoby, P. (2008). Developing a community-development approach through engaging resettling Southern Sudanese refugees within Australia. *Journal of Community Development, 43*, 483–495.

Won, N. L. L. (1999). *Guidelines on strategies and responses to the needs of Burmese migrant women in Thailand.* Asia Pacific Forum on Women, Law and Development (APWLD).

CHAPTER 9

The Power of Women Speaking Out: A Risk Assessment and Response Tool, a CSW Resolution and a UNHCR Conclusion—2005–2008

Narrated by Linda Bartolomei

A Difficult Choice

Life just kept getting busier. Attending meetings at both the Commission for the Status of Women (CSW) in New York, and UNHCR in Geneva, while juggling full-time teaching at UNSW was taking its toll, and we were getting very tired! There was a perception that the Centre for Refugee Research (CRR) was much bigger than it was, and we had not yet learned to say no to invitations to collaborate on research projects or to speak at conferences. In conversation with a colleague from a large INGO, we were asked how many staff we employed and if the university had given us a building of our own to house all of our activities! Little did he know that at that stage it was just myself, Eileen and our interns, in one room shared with ANCORW. Choices had to be made, and it was decided that we would focus on UNHCR, while our colleague Carole Shaw would lead the work in CSW.

Having survived on casual contracts since the late 1990s, I was very relieved to secure a three-year teaching contract in early 2007.[i] Teaching was central to our work and our experience in the field informed every session we taught. It allowed us to build a strong network of students and ex-students. Since taking over as the Director of CRR in 1999, Eileen had

E. Pittaway and L. A. Bartolomei,
Only Rape! Human Rights and Gender Equality for Refugee Women,
Sustainable Development Goals Series,
https://doi.org/10.1007/978-981-19-0916-0_9

provided opportunities for social work students undertaking fieldwork placements and for Master of Social Development students to intern on CRR projects. They provided essential support, self-funding their travel to assist in research consultations and training workshops. As the number of requests for training and support from refugee and community partners increased, this was formalised in 2008 by the introduction of an internship course as part of the Master of Social Development. Research associates Geraldine Doney, Effie Mitchell, Rebecca Eckert and academic colleagues Kerrie James and Eileen Baldry joined our team of volunteer trainers. Over time, a number of other academic colleagues and volunteers also donated their time and expertise and self-funded their travel to support this work.[ii]

At UNHCR Geneva, working with refugee attendees, the student team documented meetings and assisted with drafting of NGO statements. In Thailand and India, they contributed to all aspects of the work as part of a negotiated 'exchange of knowledge' with the refugee groups. They had to be willing to 'rough it', sharing beds as hard as concrete in guest houses above noisy wedding halls in the slums of New Delhi, enduring terrifying rides hanging off the cliff in the perilous streets in the town of Mizoram, the steepest city in the world, and bunking in huts in the jungle in Thailand. It was a mutual relationship. The students loved meeting and working with the refugee groups, and the refugee women and men loved the enthusiasm and sometimes irreverent fun that the students brought to their lives. We cried and laughed together in equal parts. Hundreds of students participated, and we can confidently say that there is not a place in the world where we cannot get in touch with someone we have taught!

Following the success of the Kakuma video, we produced a film as part of each research consultation we conducted. These demonstrated the power of women speaking out, the strength of their analysis and ability to contribute to solutions. The films were shown at UN meetings, in training sessions and at international conferences.

The Global Backdrop

The conflict and instability of the previous years continued. In 2005, a five-year war began in Chad and in 2006 conflicts reignited in Sri Lanka and Somalia and in the Gaza Strip. The rape and sexual abuse of women and girls continued in epidemic proportions in conflicts in Colombia, Iraq, Sudan, Chechnya, Nepal and Afghanistan (Amnesty, 2004). The war in Iraq was ongoing, and in 2007, there was an increase in the number of suicide bombings, killing hundreds of civilians. At the same time, news reports emerged of the huge number of Iraqi refugee women forced into prostitution to feed themselves and their families (Hassan, 2007). It was also a period of horrendous natural disasters. As displaced communities across Asia slowly began to recover from the devastating impacts of the 2004 Indian Ocean Tsunami, in 2005 there were new natural disasters leading to further deaths and displacement. In 2005, 1,800 people died in Hurricane Katrina and 79,000 in earthquakes in Pakistan, Iran and Indonesia. In 2008 the Wenchuan earthquake, measuring 7.8 in magnitude occurred in Sichuan, China, killing over 87,000, injuring 374,643 and leaving millions homeless. Cyclone Nargis devastated Burma and the official death toll was set at 138,366.

In 2005, there were 9.5 million refugees under UNHCR's mandate. By 2008, this would increase to over 10.5 million, as the numbers of those displaced increased due to the ongoing conflicts in Iraq and Afghanistan. Numbers of Internally Displaced People displaced by conflict grew steadily during this period to reach 26 million, with half under the protection of UNHCR (2008). With many more displaced by natural disasters, UNHCR became increasingly involved in disaster responses, following both the 2004 tsunami and the earthquake in Pakistan. In 2008, Lehman Brothers went bankrupt following the subprime mortgage crisis. The subsequent Global Financial Crisis resulted in immediate cuts in international aid, significantly increasing pressure on UNHCR and their humanitarian partners' already thinly stretched resources.

On the international stage, there were a number of significant developments in international law and policy. In 2005, the Responsibility to Protect (R2P) was unanimously adopted at the UN World Summit. It emerged as an agreed global principle which sought to ensure that the international community would never again fail to intervene to prevent genocide and ethnic cleansing as they had in Rwanda and the Former Yugoslavia (UNGA, 2005). There was also some positive progress in recognising the gendered impacts of conflict and natural disasters, to which our work had directly contributed. These included a resolution which recognised the gendered impacts of natural disasters, agreed at the Beijing Plus Ten (B+10) review in 2005, as well as Security Council Resolution 1820, which built on the commitments made in UNSC Resolution 1325 and recognises that rape is used as '*a tactic of war to humiliate, dominate, instil fear in, disperse and/or forcibly relocate civilian members of a community or ethnic group*' and must be addressed (UNSC, 2008).

Waves of Violence and Strategic Alliances

On the 27th of December 2004, the day after the devastating Indian Ocean Tsunami, we and several academic colleagues received a flurry of telephone calls from a recent graduate student, Naomi Jackson, who was in Sri Lanka. She had been at a meditation retreat in central Sri Lanka when the tsunami struck and she promptly travelled down to Colombo to offer her services to the network of humanitarian agencies who were trying to coordinate a response. The immediate need was to provide food, water and shelter. The population was in shock, and the psychological impacts of the disaster were enormous. In the first response, there was no consideration of the needs of women and girls, such as separate, secure toilets. The country representative of the UN Fund for Population and Development (UNFPA) asked if Naomi knew any international groups or organisations who might be able to assist in these areas and she contacted us and our colleague Anthony Zwi at UNSW. The university saw this as an important contribution we could make, and funded our travel to Sri Lanka. We left almost immediately.

The power of white privilege

Prior to accepting the invitation to travel to Sri Lanka, Eileen rang Sunila Abeysekera, a prominent and highly respected Sri Lankan activist with whom we had worked during the preparations for the World Conference Against Racism, in particular looking at white privilege. She had recently been awarded the Kofi Annan Human Rights Peace prize. Eileen explained that we had been funded to travel to Sri Lanka and asked whether we could make a useful contribution. Harking back to conversations they had had previously, she wryly asked, '*Do you need any more white women over there*?' '*Oh god no!!*' Sunila exclaimed. '*They are pouring off the planes*! *But if it is going to be anyone we would rather it was you, because let's face it, as white 'experts' you will be able to get into high-level meetings that we can't. And at least we can rely on you to make sure that women's voices are heard*!' And she was right, we even attended planning meetings with the President, and it would be sometime before local Sri Lankan organisations would be included in the coordination meetings convened by UN agencies and INGOs. Academics with no history of working in Sri Lanka were given precedence over a highly awarded and qualified Sri Lankan women's rights expert. The power of white privilege could not be more starkly illustrated.

Despite not being at the planning table, Sri Lankan local community organisations were the first responders, using their local knowledge to address the most urgent and immediate needs. Among them were a network of local women's organisations, who came together as The Coalition for Assisting Tsunami Affected Women (CATAW). The day after the tsunami they began fact finding missions across Sri Lanka to identify what was required in villages and towns. They also documented multiple incidents of rape, gang rape, molestation and physical abuse of women and girls in the course of unsupervised rescue operations and while in temporary shelters and displacement camps across Sri Lanka (UNFPA, 2005).

> *On the 3rd day after the tsunami a woman came to me. She told me that two men had pulled her from the waves and had asked her 'What would you prefer, to die in the waves or have sex and survive? Because of her children she decided that survival was more important, so the men raped her. She is now terrified that her husband will find out'.* (First responder, Colombo in Rees et al., 2005)

We spent a frantic and traumatising two weeks working with the Sri Lankan women's rights groups and UNFPA to assess the impact of the tsunami on women and girls and to propose responses. We also documented the high incidence of rape and sexual violence in the temporary shelters and worked with women's groups, UNFPA and with the Sri Lankan government to develop a policy for a gendered response to the disaster. There were clear parallels between the experience of women in post-disaster situations and women in post-conflict situations, and refugee camps. And once again powerful male leaders and service providers tried to deny or downplay the extent of the problem. During a debrief, the women's groups reported on a culture of denial among some displacement camp committees, medical staff, police and religious leaders. They cited comments about domestic violence, which ranged from '*this is normal*' to outright denial that any sexual violence against women was occurring. In cases where rape and sexual violence were acknowledged it was blamed on outsiders, drug addicts and criminals. Others sought to hide the extent of the sexual violence and abuse, afraid that it would impact on the international community's willingness to provide humanitarian aid.

> *The International spotlight is on Sri Lanka – people are pouring in aid. They do not need to see this disgraceful behaviour.* (Comment from a Sri Lankan doctor, in Rees et al., 2005)

The members of CATAW were so angry and distressed by what they had seen and heard, that they decided to provide some immediate training[iii] to the women community workers in the network. The participants identified that rape, sexual and gender-based violence was one of the major challenges confronting their recovery from the devastating effects of the tsunami.

> *One of the hardest things for us to understand is that in the face of such disaster, our Sri Lankan men will do this – it is breaking our hearts.* (Local relief worker, Polonnaruwa in Rees et al., 2005).

A two-day Reciprocal Research training program, *Woman to Woman*, was designed, to document barriers to safety and well-being and to involve women in developing comprehensive responses. It introduced the importance of sharing disaster and recovery stories to cope with the overwhelming distress the women were feeling. The activities reinforced the capacity and skill of women to develop appropriate responses and to protect their families and communities. The emphasis on human rights, gender equality and empowerment gave them the confidence to engage at both institutional and community levels.

An important issue in the training was to assure the women that this was not just a Sri Lankan 'problem', but in fact a universal phenomenon in the face of disasters and displacement (Enarson, 2000; Olsen, & Scharff-scher 2004). Women are central to disaster response and recovery because they are the ones with community knowledge and social networks. As with refugee populations, their voices have to be heard, and they have to be involved in all levels of decision-making, and in formal and informal leadership roles. Working with UNFPA, funding was provided to establish well-resourced women's centres across Sri Lanka, and to provide women with the training needed to run them and to collect and collate data on the progress of the recovery. These successful models were later transferred to refugee situations (Reeset al., 2005).

Advocacy at CSW for a Resolution on Gender and Disasters

For many months prior to the tsunami, women's groups across Asia had been preparing for the March Beijing Plus Ten (B+10) meeting to review further progress achieved on the commitments made in implementing the Beijing Platform for Action (BPFA). It was our final attendance at a CSW meeting. Advocates in tsunami affected countries decided that it was critical for them to raise the impact of the tsunami on women and girls at the meeting. Asia–Pacific Women's Watch (APWW) and UNFPA hosted a 'side event' and invited representatives from affected countries to present, and several international funding bodies financed the participation of women from affected areas. A representative of the Philippines Government offered to propose a resolution on the issue and members of APWW and women from the affected areas all had input into the draft and lobbied for its acceptance. The resolution was passed by consensus on the final day of the meeting. It stated that all commitments made in the BPFA were applicable to women in disaster situations, and called upon governments to integrate a gender perspective into all disaster responses and to include women in the decision-making processes (CSW, 2005: Para 2; Pittawayet al., 2007).

Developing the Women at Risk Identification Tool

Also in 2005, Erika Feller, then UNHCR Assistant High Commissioner for Protection, commissioned a trial of the Women at Risk Identification Tool in partnership with the UNHCR office and the Karen Women's Organisation (KWO) in Mae Sot on the Thai–Burma border. Two consultations were held, one in Mae La camp and the second in Mae Sot town. In a two-stage process, the refugee women first identified major risks for women and girls to be included in the Tool. In the second stage, they were joined by men from the refugee committee and NGO and UNHCR staff with whom they worked to suggest strategies to address the issues identified and to develop a series of recommendations for future action. A comprehensive list of risk factors facing refugee women and girls along the Thai–Burma Border was developed. It identified the range of human rights violations and other factors which intersect to increase the vulnerability of women and girls to violence, and which must be addressed if refugee women and girls at risk are to be provided with effective protection from all forms of violence.

Injustice and Misogyny

> *A man who committed a rape crime was charged with 2000 baht and 3 weeks in detention. He responded, 'I have 4000 baht so I can commit rape twice.'* (Refugee woman, Thailand 2005)

One of the major risk factors identified for women and girls was the lack of access to effective legal processes for cases of rape, sexual and gender-based violence. Although it was claimed that the refugee communities were entitled to seek the protection of the Thai legal system, in practice prosecutions of the perpetrators of rape, murder and other violent acts were rare. A system of camp justice operated in most of the camps, run by male refugees. Women reported that its judgements did not support punishment equal to crimes of rape and sexual violence. They were also actively excluded by male leaders from decision-making roles in most camp committees and camp justice systems. Men argued that the women did not want to take part in the procedures, and that they were not qualified to do so. The women had even been told that there were no female judges anywhere in the world. The women insisted on paralegal training and support to enable them to join the panel of judges and participate in other decision-making processes. The request was greeted with derisive laughter by male refugee committee members. More disturbing was the fact that they were joined in this derision by several of the male service providers. The 'boys club' was clearly alive and well, leaving the refugee women struggling to be heard, blocked by powerful patriarchal community leaders who were further empowered by misogynistic humanitarian workers. Fortunately, others listened, and the women received the training they requested, later joining the camp justice committees.

The research had uncovered the role of protection failures in compounding the effect of the multiple experiences of rape and abuse endured by many women and girls. The Risk Assessment Tool was expanded to include a mechanism for use by UNHCR and NGO staff, which suggested actions to be taken for each risk identified. We were invited to present this to UNHCR staff and States at the Annual Tripartite Consultations on Resettlement (ATCR) in late June (UNHCR, 2005). The women in Thailand were so determined to have their voices heard that they agreed to work with us to produce a short video of the consultations. The video *Owning Our Futures* was filmed and was sent back to Thailand for their approval before it was screened at UNHCR Geneva. Following the presentation, Erika Feller again commissioned us to work with UNHCR's Gender Unit to further develop and trial the Tool and Response Mechanism and our participatory methodology in two additional sites, New Delhi in India and a large camp in an African nation which cannot be named. Terry Morell, the Head of Women, Children and Community Services, commissioned us to develop and provide a Women's Leadership training in the camp in Africa. This was an exciting opportunity to have a significant impact on UNHCR's global policy and practice.

New Delhi: Warehoused in the City

> *One woman was raped by a group of men in the toilet of her house. Women are not safe anywhere - not even in their homes.* (Chin refugee woman, New Delhi in Pittaway & Bartolomei, 2005a)

The UNHCR Country Representative in India had requested that we undertake consultations with Chin women from Burma, to explore the risks they faced living in New Delhi. She had been concerned that recent research into risks for refugees had failed to identify and report the experience of refugee women. Instead, it had focused almost entirely on risks for refugee men. While we suspected the consultations would reveal a different story, no-one was prepared for the number and range of accounts of rape, sexual and gender-based violence documented during that five-day consultation. A shattering picture emerged of the risks of rape and discrimination faced by refugee women and girls in New Delhi in every aspect of their lives. They were not safe at work, in their homes, travelling in rickshaws, on the metro or in the marketplace. Survival sex was the only way many women survived and protected their children.

Total impunity
Perhaps one of the most disturbing stories the women shared was of being forced to fight with cows and crows for vegetable scraps discarded each night by vegetable sellers in the market, who would also sometimes urinate on the vegetables in an attempt to prevent the refugee women from collecting them. We made the mistake one night of going to the markets with a male colleague from Amnesty International, who wished to take photographs to raise awareness of the issue. On seeing us, one stall holder reached over and roughly fondled the breast of a refugee women, while looking at our friend and saying *'Is this what you want to see? Try and stop me!'*.

UNHCR took immediate steps to address the problems, instigating monthly meetings with the refugees and opening a safe space for the women, which provided emergency aid, counselling and support for those who had experienced rape, sexual and gender-based violence. This was funded by the Australian Government. It was the first time we had fully documented the extent of the abuse experienced by women in urban settings, challenging the growing sentiment among humanitarian aid workers that refugees were in fact safer in cities than in camps. We noted this was something that we needed to follow up as soon as we could.

And then We Went to Africa[IV]...

> *The men come to our houses at night to sleep with us – to rape us and our daughters. There is no-one to protect us and the government people and UNHCR do nothing when we tell them this happens. Men wait for us when we go to the market - a man hid in a tree and raped me in the daytime when I was going to the market. What kind of governance do we have in the camp when this happens to women all the time?* (Refugee woman, Africa in Pittaway & Bartolomei, 2005b)

The camp, which cannot be named for reasons of confidentiality (Pittaway & Bartolomei, 2005b), is in a remote border location and was home to 35,000 refugees from five different ethnic groups who had fled conflict in a neighbouring country. There were also a small number of refugees from several other African nations, who had recently been relocated from the capital. In direct contrast to the majority of refugees they were highly educated and had come from urban areas in their home countries. This difference significantly increased the risks they faced in the camp as they were culturally and linguistically very different to those from rural areas and there was antagonism between the groups. We spent two weeks in the camp, using Reciprocal Research and conducting women's leadership training. However, what started as a training exercise escalated into an expose of some of the worst abuse, corruption, neglect and multiple failures of protection we had yet witnessed.

Researchers, Trainers or UN Investigators?

> *We are so pleased that you are women who have come to do this training – this is a place where everything is kept secret. It is good that at last we can speak out.* (Refugee woman, Africa in Pittaway & Bartolomei, 2005b)

In all zones of the camp and across ethnic groups, widows and single women reported rape and continual sexual abuse. There were serious problems in food distribution: women reported receiving only 13 of their 15 kilogram grain allocation per month, and seeing trucks removing food in the night. The small amount of food received plus the inability to procure fresh food items contributed to malnutrition in the camp and as a result some children were being tube fed in the 'clinic' (Fig. 9.1).

> *I have not drawn these people as stick figures because I cannot draw but to show you how thin they are because they are starving.* (Refugee woman 2005)

Fig. 9.1 Storyboard—Malnutrition, 2005

A representative from the World Food Program (WFP) visited the camp when we were there and expressed his disgust at what was happening, stating he had never been informed of this. He committed to an urgent local investigation. The government organisation that ran the camps and had responsibility to distribute food denied the allegations made by the refugees. Women also reported that they only received one reusable sanitary towel each year, made from non-absorbent material, and no extra soap to clean them. When this was disputed, they explained through a male interpreter that if a man had a cut on his leg and he tried to use a sanitary towel to the staunch the bleeding, the blood would just run under the material. The lack of food security and non-food items forced many women to sell sex to procure such basic items as sanitary napkins. They were also targeted for forced prostitution or coerced into 'marriage' by much older men. Several UN and NGO staff kept young female refugees as live-in housekeepers or 'wives'.

The ultimate love for a child

One of the very young single women in the training had a baby girl who was obviously the centre of her life. Each day she brought her to the training, and I would play with the baby, holding her while the woman was working. On the final day, the woman brought the baby, beautifully dressed, and handed her to me. '*You take her to Australia*' she said. '*There is nothing for her here, no future, no hope. She will grow up like me*'. With tears in my eyes, I explained that this was not possible, and that the baby needed her mother and the love she was giving her. '*No*' the mother replied, '*she would have a future with you*'. I was not the only one crying as we left.

Serious issues were also identified in the health clinics, including lack of adequate medicine and insufficient and poorly trained staff. There were few if any female paramedics, and rape victims who did report were not allowed to be accompanied by a friend for their examination. Many of the women complained that the only medication distributed was paracetamol, regardless of the symptoms. The male clinic manager informed us that this was a technical matter: that the women were so ignorant they did not realise that these days one paracetamol could cure many diseases and that they did not need a range of medications. As the women from the rural and urban backgrounds shared stories and realised the similarity of their experiences, barriers broke down, and before we left informal support structures were established between the groups.

The stated aim of our visit was to provide leadership training to the women, and we were told several times in the first couple of days that women were included in all committees and that they were being untruthful when they said they were not. Towards the end of the visit, it was acknowledged that they had no place nor space in the decision-making process, and the chief religious leader was totally dismissive of the suggestion that they should. It was apparent that the major power brokers were the government officials and the male community leaders. The only way to communicate with these bodies was in writing and through the camp committee, but the women were mainly pre-literate. They described how the gates to UNHCR and the government offices are always locked, and if they knock, they are never allowed to enter. Women were very aware of their potential, their role and responsibility in the camp, but identified that there are no resources to support them to take leadership.

There was a constant reference to the culture of refugee communities as an excuse for the lack of protection, but it was women from the same cultures who were raising the concerns.

> *We constantly hear them [service providers] telling visitors that we are so backward that we arrived in the camp naked and they had to find us clothes, that we have to be looked after like children. They never talk to us, or they would have heard about how our journey was so long and terrible that our clothes turned to rags and became full of lice. We had to burn them because our skin was infected. But that is how they see us and treat us. Can you imagine how it feels?* (Refugee woman, Africa 2005)

When the issues raised by the women were shared with government and UN service providers in a preliminary debriefing session, we were advised that a functioning sexual and gender-based violence committee and response mechanism had been established. Investigation revealed that despite the election of a community-based committee composed of six women and six men, plus zone representatives, 20 months earlier, it had not once met since it was formed. The UNHCR Protection Officer indicated that refugee women will often exaggerate claims of sexual and gender-based violence to obtain resettlement, yet despite this, in 42 hours of training with 49 women, resettlement was only briefly mentioned twice, and then as a potential solution for widows and single women at extreme risk. There was a constant characterisation of refugees as manipulative, liars and cheats, manufacturing or exaggerating stories to obtain resettlement.

Another 'Royal Visit'

To add insult to injury, while we were there, the UNHCR High Commissioner made a short visit to the camp, and we witnessed the amount of time, money and resources spent on this event. Staff reported that the preparation had taken 'centre stage' for the previous two months, often at the expense of protection activities. It included installing an outdoor tap and hand basin so he could wash his hands before lunch in a camp where there was a chronic shortage of bathing facilities and water points. The normally ragged children who met him were given new tee-shirts. The most obvious abuse of refugee rights was the gallons of water used to damp down dust in places the High Commissioner might visit. This, in a site where water was an extremely scarce resource and refugee women and children had to carry their limited supplies of water in jerry cans from distribution points for their every need. These expensive preparations were undertaken in face of the often-stated fact of desperate budget and staff resource constraints.

The women we were working with had prepared a short presentation for this visit and were promised he would come to hear it. To meet him they stayed for hours in an exceedingly hot tin-roofed room without lunch or water. He did not visit, nor did any senior UNHCR staff show the women the courtesy of visiting the training venue to advise them that he did not have sufficient time to meet them. Food was purchased by us for the women, but water could not be provided as the nearest tap was not switched on for the daily two hours of water allocation during this period. This was the final day of a week-long training during which the women had continuously expressed their concern about the fact that their voice was never heard. They were extremely vocal in their disappointment and saw this as a further example of their marginalisation in the power structure in the camp. Several questioned the value of a training program financed by UNHCR Geneva aimed at engendering leadership qualities in refugee women, when the High Commissioner did not even stop by for a couple of minutes when he had been in a hut adjacent to the one in which they were waiting.

On leaving the camp we attended a debriefing in the capital city in which one very senior member of UNHCR dismissed the concerns of women on the basis that all refugees lie and exaggerate to visitors and that some of the problems which we reported did not exist. However, this was directly contradicted by senior NGO and UN staff in the same meeting. While clearly displeased that we were shining a spotlight on the serious protection failures in the camp, they reluctantly acknowledged that they had been aware of the problems for over a year. They emphasised that the ongoing shortfall in international humanitarian aid funding had led to a severe lack of resources, staff and training which was contributing to the problems. Towards the end of the meeting, it became almost farcical as it descended into a game of one upmanship, as they tried to imply that they in fact knew of things far worse than those we had observed and heard and that the camp we had visited was not the worst one in the country. This begged the question of why so little action had been taken.

At the end of this visit, we were in a very difficult situation. We did not go to the camp as an investigation team, but that is how it had turned out. We had no mandate, other than to undertake risk assessment from the perspective of the refugee women, but they had disclosed issues which were potentially criminal, which certainly contravened the UNHCR Code of Conduct and were human rights abuses. We produced two reports, one about our work with the women, and a second confidential report to UNHCR senior management. In this, while we acknowledged the serious budget, resource and staff constraints faced by UNHCR and many of the NGO Implementing Partners, we fully detailed the numerous protection failures identified during the consultation and training held in the camp. As serious allegations had been raised by refugees against the local government authorities, as well as the camp service providers, we held serious concerns that there would be retributions after we left the camp. It was imperative that the report be received in UNHCR Geneva as quickly as possible, and we sent it five days after arriving home, on December 26th.

The response from UNHCR was far ranging, and actions were taken at all levels to improve the situation. Investigation found that the evidence provided by the women was not in fact 'lies and exaggeration', and changes were made to service provision in the camps. It was found that many women had a valid claim to be resettled as women at risk, and this occurred.

What we learned from this experience was how totally divorced headquarters in Geneva was from the field, and the power which this gave to 'street level bureaucrats' on the ground. The chronic lack of funding meant that there were not sufficient resources to effectively support staff to operate effective accountability mechanisms and lines of reporting were extremely weak. It also demonstrated the power of local authorities in host countries to determine the level of protection provided. While satisfied with the prompt response, we were left questioning if we had in fact been sent to this camp not just to undertake training, but in the hope we would be the catalyst for change. Although 2005 had been bookended by the horrors of the tsunami at the start and our findings in the refugee camp at the end, we remained hopeful that our direct work with the women's groups had provided them a platform to have their voices heard.

Back at UNHCR for the 2005 ExCom meeting

We returned to Geneva in June for the annual Pre-ExCom and ExCom meetings with the Australian Refugee Rights Alliance (ARRA), including ANCORW members Olivia Wellesley-Cole, from Sierra Leone and Juliana Nkrumah, from Ghana, both well-respected advocates for the rights of refugee women, and several other refugee women advocates. While at the meeting, the *Refugee Women at Risk Assessment Tool and Response Mechanism* (Pittaway & Bartolomei, 2005c) was launched.

Despite our advocacy for a Conclusion on Women at Risk (WaR), we had been advised that we would not be invited to take part in drafting one, because '*NGOs and academics don't draft conclusions*'. The usual process is that conclusions are drafted by UNHCR and then modified based on the suggestions and input of member states (Sztucki, 1998). However, following sustained lobbying, at the previous December Standing Committee, Erika Feller had turned to Eileen and said '*OK, give me a draft and I will see what we can do*'. We quickly produced a draft, based on the key findings of our research. *The Case for the Conclusion on Women and Girls at Risk* (Pittaway & Bartolomei, 2005d) presented a brief background and justification and suggested text using the style of language used in UNHCR Conclusions. The proposed Conclusion aimed to provide guidance on addressing the identified implementation failures through the provision of better targeted guidelines, monitoring mechanisms and resources. It was taken forward into the long slow process of UN decision-making and formally launched at the June meeting with the support of refugee women who had contributed to its development.

Meanwhile in Australia

We continued to work with ANCORW and other partners across Australia, providing training when requested and working on joint advocacy strategies to improve settlement services and the implementation of the WaR Program. There were many reports of family and domestic violence in refugee communities, and we received funding to explore links between pre-arrival and post-arrival experience, of grief, loss, torture and trauma. Olivia Wellesley -Cole, an ANCORW board member from Sierra Leone, worked with refugee communities to analyse the problems they were experiencing and suggest ways to improve service provision. Models of response were then developed and trialled with the communities and a training kit and film titled *Horror to Hope*[v] was produced. While we have reported on our work overseas separately from that in Australia, in fact they are two sides of one coin. In talking with women resettled in the relative safety of Australia, they felt free to tell us more of things that had happened in refugee sites, and we were able to use this in our work overseas. In fact we discovered that if we did not give women space to talk about their pre-arrival experience, they were often reluctant to talk about life in Australia. In return we were able to use the gains that we made at UNHCR to inform our advocacy and work in Australia.

Refugee Women at Risk: Protection and Integration in Australia, ARC Linkage Project

> *What about single women? They are so much at risk. They struggle every day to resist rape. People know you are alone, and men try sexual abuse.* (Resettled refugee woman 2008 in Bartolomei et al., 2014)

In close collaboration with service providers in Townsville, Wagga Wagga and Wollongong, which provided settlement services for newly arrived refugees, we learned that the WaR Program was still not functioning effectively. While numbers had steadily increased, funding to provide the additional services they required had not. Women and their families were not settling well; they were experiencing multiple and severe problems and discrimination, both from the Australian and their own communities. Single women with children who were known to have arrived on a WaR Visa were assumed by others in their communities to have been raped or to have sold sex to survive. They reported being shamed, and sometimes threatened and abused by men who knew of their history. Far from being a safe haven and a new secure home, Australia was yet another site of fear and marginalisation.

In 2006, with colleague Professor Richard Hugman, we applied for Australian Research Council (ARC) funding with ANCORW to see what could be done to address this. Along with PhD student Rebecca Eckert, who had been volunteering and working with us for two years, and Effie Mitchell, we commenced the three-year-long project in 2007. This project included a particular focus on regional settlement in Australia. It built on the outcomes of the previous ARC-funded project and the outcomes of an International Conference on Resettlement, *Hopes Fulfilled or Dreams Shattered*, that we had hosted in November 2005. These had identified key settlement challenges for women at risk and their families which need to be urgently addressed.

It was incredibly challenging once again hearing the stories of women who had endured rape, sexual torture and severe discrimination, and who continued to do so in Australia. The accounts of racist behaviour displayed by some service providers and the Australian community made us furiously angry and sick with shame. Fortunately there were also amazing members of refugee communities, services providers and volunteer support groups who really cared and did everything they could to ease the passage of the women into their new homeland. The women who had arrived on WaR Visas did not only have sad tales to tell. They too analysed what was happening to them, identified what was really helping them in Australia and made many suggestions for changes. These were taken up by advocacy groups round Australia, and many very responsive settlement service providers were able to amend their practice and provide more comprehensive services. A major shift at this time was that funding was made available for the employment of women from refugee backgrounds as multilingual workers in settlement services.

Advocacy for the Conclusion—ExCom 2006

Getting the WaR Conclusion adopted by ExCom required some of the most persistent lobbying we had done yet, because there was still considerable resistance to acknowledging the extent of the issue. At one UN meeting we were told so often that 'women will not talk about rape' that we quickly cobbled together and presented a three-minute video from footage on our cameras of women from around the world doing just that. It certainly silenced critics on that occasion. Without the support of key staff in UNHCR and the sustained advocacy of several Geneva-based NGOs, in particular the Jesuit Refugee Service (JRS), Women's Commission for Refugee Women and Children (WCRWC), the Quakers International, Amnesty International, the International Catholic Migration Commission (ICMC) and the International Council of Voluntary Agencies (ICVA), it would not have succeeded. Their presence in Geneva enabled them to maintain sustained pressure on key governments to support a Conclusion on WaR. Australia has never felt so far from the seat of power! They kept us informed of the status of negotiations and shared with us the draft text of the Conclusion for comment as it developed. Christina Bloch, JRS Mariette Grange, ICMC and Manisha Thomas, ICVA, were particularly strong allies. They would find us wherever we were in the world at any hour to update us on the progress of the negotiations. One particularly memorable call came while we were leading a training at the American University in Cairo on the invitation of the legendary Barbara Harrell-Bond. Christine was fighting to retain the word 'rape' in the text, but one government was insisting it be removed. Despite her efforts, the government representative would not budge and as she apologetically explained when she called us later that night, '*deleting the word rape was the only way that the Conclusion could go forward.*' It once again demonstrated the enormity of the ideological and political struggle in the fight for women's human rights.

The adoption of the *Conclusion on Women and Girls at Risk* on the 6th October 2006, by UNHCR's Executive Committee was an important and impactful milestone in this struggle. The contribution of our research to this Conclusion was publicly acknowledged by both Peter Hughes representing the Australian Government (Australia 2006) and by Erika Feller, UNHCR Assistant High Commissioner for Protection (ARRA, 2006). Following its adoption, UNHCR's then High Commissioner, António Guterres, dedicated $25 million from his discretionary fund for protection programs for women and girls (ARRA, 2006). Our research and advocacy led to significant and sustained increases in the numbers of refugee women and girls at risk identified by UNHCR and referred for resettlement to countries including Australia.

What was unique about the Conclusion and The Heightened Risk Assessment Tool was the role that refugee women themselves played in their development. The text of both these documents reflected the input of refugee women's groups we had worked with in Kenya, Thailand, India and Australia over the previous three years. We continued to assist the women in their own advocacy by editing and distributing their own reports and facilitating their participation in international meetings including UNHCR and the B+10 meeting (KWO, 2007; WLB, 2005). In 2006, the year the Conclusion was formally adopted, we were able to fund one of our refugee women partners, Tenneh Kpaka, to attend and speak at the UNHCR ExCom meeting. A refugee woman formally from Sierra Leone and on the board of ANCORW, she was one of the first refugees to present the formal NGO statement to ExCom. As researchers we strove to stay accountable to the refugee women and provide entrée to the channels to which we had privileged access. By 2006 the Australian Government along with several other governments had publicly announced their support for improved implementation of the WaR Program and other programs to improve protection for refugee women and girls (Australian Government, 2006).

Implementing the Conclusion on Refugee Women at Risk

In March 2007 we were invited by UNHCR to lead a 'multi-functional' team to formally trial the Tool and Reciprocal Research methodology in the Rohingya Refugee Camps in Cox's Bazar, Bangladesh. This time it was openly acknowledged that there were very serious problems in the camps, and that it was extremely difficult to engage the women because they were very conservative Muslims who did not like speaking with strangers! The team included Senior UNHCR policy staff from Geneva, INGO AUSTCARE and staff from Foundation House in Melbourne, a Torture and Trauma service provider who also worked closely with UNHCR. Graham Thom from Amnesty International joined the UNSW team. Prior to the trial we worked with staff from Foundation House in Melbourne to expand the Tool from its sole focus on women and girls at risk to include other at-risk groups. These included survivors of severe torture and trauma, unaccompanied minors, refugees with a disability and aged refugees. In Cox's Bazar we first provided training to the team on the use of the Reciprocal Community Consultation method we had developed. We then spent ten days in the Kuttapalong and Nayapara camps undertaking four separate community consultations, and individual interviews with women and men. As always, we provided each of the consultation participants in the consultations with a small bag of multicoloured beads to make a symbolic human rights necklace.

Difficult as it was to believe, the consultation and interview reports detailed some of the most brutal and violent human rights abuses we had yet documented. A network of dishonest refugee leaders linked to the local village mafia and corrupt local officials had for many years presided over a brutalised and terrorised refugee population. Rape, survival sex, and the trafficking of girls and young children into prostitution were endemic. Men were beaten, tortured and frequently arrested and imprisoned on false charges if they dared to protect the women, or to challenge the corrupt leaders. Food rations were stolen and sold on the black markets; young people were denied the right to education or even the space to play (UNHCR, 2007) (Fig. 9.2).

> *Inside the camp often when women go for water they are then taken and raped*'. (Rohingya refugee woman 2007)

A tool adapted from a Harvard University instrument designed to measure trauma was trialled by colleagues investigating trauma in the population. The scale included questions about difficulty sleeping and loss of appetite. It provided a stark example of the lack of context-appropriate tools and the severe limitations of western models. Given the visibly

Fig. 9.2 Storyboard—Rape of Rohingya women, Kutupalong Camp 2007

appalling conditions in the camp, several of the Rohingya male refugees who spoke English shook their heads in disbelief and asked if we were mad.

At the end of the consultations, the participants nominated refugees to report back on the findings to UNHCR and other stakeholders. People came from UNHCR in Geneva and Dhaka, from all of the NGOs, and the local authorities to listen to what the refugees had to say. We had separate presentations in each camp. In Kuttapalong the women nominated an older woman to make the presentation. She was not educated but she was their leader. They chose the storyboards for her to talk about and discussed among themselves the most important messages they wanted to be heard. The presentation took place in a huge hut: all of the 'important' people sat up on a stage and everyone else sat on the floor, women to one side, the men to the other. It was an incredibly poor camp, with severe water restrictions. All of the women wore black burkas, torn and dusty.

Why we teach human rights

The older woman was small but she stood tall facing the stage and the interpreter held the microphone for her. She started like this. *'Good morning UNHCR and NGOs. I have pictures here ready to show about what we have discussed this week, but first, please look at me and you will see that over my burka I have a human rights necklace'*. People looked surprised and confused; they wondered why she was talking about a necklace. Then she said, *'Look at the women, they all have the necklace on'* and we looked and all of the women had their brightly coloured beaded necklaces on the outsides of their dusty, black burkas. She then said, *'Look at the men, they are all wearing necklaces'* and every man had also put on a necklace. She said, *'UNHCR and NGOs let me tell you about my necklace. These stand for all of our human rights. When I was a girl, back in my village in Burma, we had most of these rights. We had happy families, we went to school, we had enough food, we had water. We didn't have all of the rights but we had a lot of them. Then the war came, we were persecuted, the military burnt our villages, raped the women, beat-up and killed the men. Horror came and we had to flee to Bangladesh. But we ran with our human rights necklaces on our necks because we believed that in Bangladesh, we would have safety and we would have our human rights.* She paused and looked hard at the dignitaries and said, *'UNHCR and NGOs, this camp is terrible, it is dangerous and we are starving. Do you know what is happening to our human rights here?'* Then she lifted her hands to her neck and tore off her necklace. She pointed to the beads scattered across the floor and said, *'This is what has happened to our human rights. UNHCR and NGOs, it is your duty to help us to pick up them up from the floor. We all want to wear our necklaces again.'* She then picked up her storyboard pictures and gave the presentation that she had done with the women (recounted by Eileen Pittaway, 2007).

That night when we debriefed, Eileen shared the story and told us that it had been one of the most powerful and wonderful things she had witnessed. The report and a short film from these community consultations, approved by the participants before their release, led to direct and immediate improvements in the camps' environment and in the protection response of UNHCR and governments. Following the formal launch of the report in mid-2007, a number of major donor countries including Australia made significant financial contributions to improve the living conditions. The Bangladeshi government responded positively by allowing in a number of international NGOs who had previously been refused permission to work in the camps. Most importantly, where resettlement had not previously been an option, a number of governments agreed to open their programs to refugees at heightened risk. A group in the UK saw the film and used it to lobby their government to open up resettlement in England for Rohingya refugees. By 2008, over 600 places had been offered for resettlement by the governments of Canada, Australia, USA, the UK and Ireland. The revised and expanded Tool was launched as the UNHCR Heightened Risk Identification Tool (HRIT) in June 2008. It is now used as part of the Standard Operating Procedures in UNHCR operations around the world. The women's human rights necklace story is an integral part of Reciprocal Research training.

Further Expanding Our Advocacy Networks

Our links with other groups were strengthened as we strived to contribute to transnational advocacy networks. These included the Southern Refugee Legal Aid Network (SRLAN), bringing together refugee rights NGOs, academics and lawyers from across Africa and the Middle East (Harrell-Bond, 2007). Inspired by the power and success of this network to share information and network for advocacy, Alice Nah, a refugee rights advocate based in Malaysia, began to canvas interest among NGOs and academics based in the Asia region to form a similar network in the Asia–Pacific. In November 2008, with the support of Forum Asia,[vi] she hosted an inaugural planning meeting in Kuala Lumpur. Effie Mitchell and I participated, becoming founding members of the Asia–Pacific Refugee Rights Network (APRRN). The structure included sub-regional geographic and thematic working groups and we proposed a thematic working group on Refugee Women at Risk which was endorsed by the meeting. I was elected the inaugural working group chair with Katrina Maliamauv from the Malaysian migrant rights NGO Tenaganita elected as deputy. This marked the beginning of a long and mutually fruitful relationship with APRRN and Tenaganita.

Listen to the Women!

Our relationship with WLB continued. Twice a year we would be on a plane heading to either India or Thailand to provide trainings. The choice of training was totally in the hands of the women, and topics ranged from new Security Council resolutions relating to women and peacebuilding, feminist human rights documentation and accidental counselling. The one time we privately questioned the topic was when they requested training on 'Responsibility to Protect' (R2P). We wondered how they would use the knowledge? then, reminding ourselves of our promise to them, we prepared and delivered the training. Soon afterwards, some of the women from the training travelled with us to a meeting at Geneva, where we introduced them to people we knew on various government delegations.

Fast forward a couple of years to Cyclone Nargis. The world looked on in horror as despite international pressure in the Security Council, the Government of Myanmar refused to allow foreign aid to the desperate and starving survivors, many of whom were from the same ethnic groups as the refugees we worked with. As the ethnic communities were particularly affected, grassroots women's groups, including WLB and KWO, mobilised to advocate for international action and support.

At midnight one night, we received a phone call from a very excited refugee woman leader. '*Listen to the news and we will talk tomorrow!*' was all she said. As we listened to the news, it was announced that the Association of Southeast Asian Nations (ASEAN), had persuaded Myanmar to allow aid to enter and it would start arriving the following day. When we finally spoke she told of how she and her group had approached the

French Ambassador, whom they had met in Geneva, and requested him to evoke R2P in the Security Council. He agreed and although it was vetoed by China and several other member states, the process appeared to have alerted ASEAN members to the seriousness of the situation. They agreed to put their weight behind the push to have Myanmar open up to foreign aid. The women had clearly demonstrated that they had grasped the fine detail of international politics even more than we had! No wonder so many nations did not want them further empowered!

Lessons Learned

Our analysis of the work we undertook in this period and the evidence we collected confirmed for us the extent, complexity and the depth of the problems we were investigating. Rights-based participatory research, on which Reciprocal Research is grounded, had proved to be a very effective tool to challenge the silences we were uncovering. These included the silences of humanitarian actors complicit in abuses, and of powerful men in refugee communities who, by denying women access to a voice and justice, strove to keep the violence and abuses hidden. Our research had continued to critically unpack the range of political and ideological barriers in both the policy framework and its implementation which not only inhibited the effective protection of refugee women and girls but often only exacerbated the risks they faced. Following Turton (2003, p. 2), we had used our rigorous academic research and critical policy analysis 'not to sustain and legitimise the practical knowledge upon which policy is based, but to scrutinise and problematise it.' This was most evident in our constant challenging of the myths, assumptions and patriarchal attitudes which underpinned so many of the protection failures we had documented. Courageous refugee women and girls had once again spoken out in camps and cities across the world. The passing of the UNHCR Conclusion on Women and Girls at Risk in 2006 and the adoption of the Heightened Risk Identification Tool as standard operating practice in 2007, had ensured that they were not just heard, but that their lived experience had directly contributed to significant changes in law, policy and practice.

Notes

i. With special thanks to Professors Jan Breckenridge, Richard Hugman and Eileen Baldry!
ii. This included, among others, now Professor Eileen Baldry, Graham Thom, Emma Pittaway, Tashi James, Colleen Bartolomei, Meg Davis and Connie Hensen.
iii. Financial support was provided by UNFPA and AUSTCARE.
iv. Juliana Nkrumah joined us as a trainer, as did UNHCR Gender Advisors from Geneva and the Regional Office in Nairobi.
v. Funded by the NSW Department for Women.
vi. Funding was provided by UNHCR, AUSAID, the High Commission of Canada in Kuala Lumpur, Inter Pares, Euro-Burma Office, and the International Detention Coalition.

References

Amnesty International. (2004). *Lives blown apart: Crimes against women in times of conflict*. Amnesty International. Retrieved November 10, 2021, from https://www.amnesty.org/en/wp-content/uploads/2021/08/act770752004en.pdf

Australian Government. (2006). *Australian Intervention, UNHCR Executive Committee October 2006 AGENDA ITEM 5a, International Protection*. Centre for Refugee Research, Sydney. Retrieved July 14, 2009, from http://www.crr.unsw.edu.au/documents/Australian%20Intervention%20on%20Protection.pdf

Australian Refugee Rights Alliance (ARRA). (2006). *Report of the 57th session of UNHCR Executive Committee's annual consultations with non-governmental organisations September 2006*. Refugee Council of Australia, Sydney. Retrieved October 15, 2009, from http://www.refugeecouncil.org.au/docs/resources/reports/ARRArpt_06.pdf

Bartolomei, L., Eckert, R., & Pittaway, E. (2014). "What happens there ... follows us here": Resettled but still at risk: Refugee women and girls in Australia. *Refuge: Canada's Journal on Refugees, 30*(2), 45–56. https://doi.org/10.25071/1920-7336.3961

BBC. (2008, June 20). UN classifies rape a 'war tactic'. *BBC News*. http://news.bbc.co.uk/2/hi/americas/7464462.stm

Commission on the Status of Women (CSW). (2005). *Forty-ninth session 28 February-11 March, 2005 Item 3(a) of the Provisional Agenda*. United Nations. Retrieved November 10, 2021, from https://www.un.org/womenwatch/daw/csw/csw49/documents/E-CN-6-2005-CRP1-E.pdf

Enarson, E. (2000). *Gender issues in natural disasters: Talking points and research needs*. ILO InFocus Programme on Crisis Response and Reconstruction Workshop.

Harrell-Bond, B. (2007). Starting a movement of refugee legal aid organizations in the South. *International Journal of Refugee Law, 19*(4), 729–735. https://doi.org/10.1093/ijrl/eem063

Hassan, N. (2007, June 24). '50,000 Iraqi refugees' forced into prostitution. *The Independent*. https://web.archive.org/web/20080708215548/http:/news.independent.co.uk/world/middle_east/article2701324.ece

Karen Women's Organisation (KWO). (2007). *State of terror*. KWO.

Olsen, O., & Scharffscher, K. (2004). Rape in refugee camps as organisation failures. *International Journal of Human Rights., 8*(4), 377–397.

Pittaway, E., & Bartolomei, L. (2005a). *New Delhi: Report to United Nations High Commissioner for Refugees*. Centre for Refugee Research. On file with author.

Pittaway, E., & Bartolomei, L. (2005b). *Deidentified report to United Nations High Commissioner for Refugees*. Centre for Refugee Research. On file with author.

Pittaway, E., & Bartolomei, L. (2005c). *Refugee women at risk assessment tool and response mechanism*. Centre for Refugee Research / Australian National Council on Refugee Women. Retrieved October 15, 2009, from www.crr.unsw.edu.au

Pittaway, E., & Bartolomei, L. (2005d). *The case for a conclusion on refugee women at risk*. Centre for Refugee Research / Australian National Committee on Refugee Women. Retrieved October 15, 2009, from www.crr.unsw.edu.au

Pittaway, E., Bartolomei, L., & Rees, S. (2007). Gendered dimensions of the 2004 tsunami and a potential social work response in post-disaster situations. *International Social Work, 50*(3), 307–319.

Rees, S., Pittaway, E., & Bartolomei, L. A. (2005). Waves of violence—Women in post-tsunami Sri Lanka. *Australasian Journal of Disaster and Trauma Studies, 2*, 1–6.

Sztucki, J. (1998). The conclusions on the international protection of refugees adopted by the executive committee of the UNHCR programme. *International Journal of Refugee Law, 1*(3), 285–318.

Turton, D. (2003). Refugees, forced resettlers and 'other forced migrants': Towards a unitary study of forced migration. *New Issues in Refugee Research, Working Paper No. 94*, Evaluation and Policy Unit UNHCR. https://www.unhcr.org/en-au/research/working/3f818a4d4/refugees-forced-resettlers-other-forced-migrants-towards-unitary-study.html

United Nations High Commissioner for Refugees (UNHCR). (2005). *Report on the annual tripartite consultations on resettlement, Geneva*. Retrieved June 10, 2007, from www.unchr.org

United Nations High Commissioner for Refugees (UNHCR). (2006). *Executive Committee Conclusions, No. 105 (LVII), conclusion on women and girls at risk.* UNHCR. Retrieved November 10, 2021, from https://www.unhcr.org/en-au/excom/exconc/45339d922/conclusion-women-girls-risk.html

United Nations High Commissioner for Refugees (UNHCR). (2007). *Refugee consultations—Bangladesh, March 2007*. UNHCR.

United Nations High Commissioner for Refugees (UNHCR). (2008). *2008 global trends: Refugees, asylum-seekers, returnees, internally displaced and stateless persons*. UNHCR. Retrieved November 10, 2021, from https://www.unhcr.org/statistics/country/4a375c426/2008-global-trends-refugees-asylum-seekers-returnees-internally-displaced.html

United Nations General Assembly (UNGA). (2005). *Resolution adopted by the general assembly on 16 September 2005*. United Nations. Retrieved November 10, 2021, from https://www.un.org/en/ga/search/view_doc.asp?symbol=A/RES/60/1

United Nations Population Fund (UNFPA). (2005, January 5). *UNFPA calls for greater security for women affected by tsunami*. United Nations Population Fund Press Release. Retrieved November 10, 2021, from https://www.unfpa.org/press/unfpa-calls-greater-security-women-affected-tsunami

United Nations Security Council (UNSC). (2008). *Resolution 1820, adopted by the Security Council at its 5916th meeting, on 19 June 2008, S/RES/1820.* United Nations. Retrieved November 20, 2008, from http://www.un.org/Docs/sc/unsc_resolutions08.htm

Women's League of Burma (WLB). (2005). *Any progress for the lives of women in Burma since Beijing?* WLB.

CHAPTER 10

Survivors, Protectors, Providers: Refugee Women Speak Out from Gender Mainstreaming to the Regional Dialogues—2009–2011

Narrated by Linda Bartolomei

When women get the chance to be heard - the world changes! (Refugee woman leader, Uganda 2011)

Age, Gender and Diversity Mainstreaming

Because of our unique position travelling between refugee sites and the United Nations, often as keynote speakers and experts at UN meetings, funding to attend conferences increased. This enabled us to debate concepts such as whether rape in conflict was torture and to widely distribute our advocacy films and reports. One of the significant policy changes undertaken by UNHCR in this period was the development of the Age, Gender and Diversity Mainstreaming Strategy (AGDM), which aimed to facilitate 'targeted action to address inequalities and support the empowerment of discriminated groups, in particular women and girls at risk' (Okoth-Obbo, 2009, p. 1). In spite of the genuine efforts of some very senior and committed UNHCR staff to keep the needs of women and girls centred, and to emphasise the importance and value of this approach, it was increasingly evident that this did not translate into policy or practice on the ground. 'When we no longer want to dedicate

E. Pittaway and L. A. Bartolomei,
Only Rape! Human Rights and Gender Equality for Refugee Women,
Sustainable Development Goals Series,
https://doi.org/10.1007/978-981-19-0916-0_10

specific resources to a group, we either mainstream it or we call it a cross-cutting issue' (UN Policy-maker, Ferris, 2014). Rather than facilitating the increased visibility of gender, it had the opposite effect. Instead of being recognised as more than 50% of the refugee population with diverse backgrounds, needs and experiences, once again, women were lost within the discourse of 'vulnerable groups'.

Meanwhile, the World Continued Its Chaotic Trajectory

In 2009 the Sri Lankan Civil War ended although sporadic fighting continued. The World Health Organization declared 'swine flu' a global pandemic. The European Union signed the Lisbon Treaty establishing a more centralised leadership and foreign policy, threatening tighter restrictions on asylum seekers. Barak Obama was inaugurated as 44th President of the USA generating much hope for a more secure and democratic future. In 2010, a military crackdown in Thailand targeted the United Front for Democracy against Dictatorship (UDD). North Korea shelled South Korean forces stationed on Yeonpyeong Island. The so-called Arab Spring began with a series of pro-democracy protests, uprisings, and armed rebellions that spread across much of the Arab world in the early 2010s, and, in Nigeria, there was a coup d'état. Meanwhile, the sovereign debt crisis faced by Greece in the aftermath of the financial crisis of 2007–2008 caused major hardships to the Greek population and the global economy.

In 2011 the Republic of South Sudan gained independence from the Republic of the Sudan bringing short-lived hope of an end to conflict and forced migration in that region. The Arab Spring continued with major conflicts in Tunisia, Egypt and Libya. In Syria, civil war broke out between the Ba'athist Syrian Arab Republic, led by President Bashar al-Assad, and various domestic and foreign forces opposing both the Syrian government and each other in varying combinations. The Second Ivorian Civil War began. On a positive note, the 'Occupy movement', an international movement expressing opposition to social and economic inequality and the lack of 'real democracy', inspired worldwide protest. The withdrawal of US military forces from Iraq was completed by December 2011, ending the Iraq War.

World population reached seven billion, with many living in abject poverty, yet data by country indicated that global inequality was falling.

Credit Suisse, the International Banking Corporation, estimated that in 2011, 0.06% of the world's population owned 40% of the world's wealth, while the bottom 95% shared only 28%. Conflicts and inequality caused ever-growing numbers of people to be forcibly displaced. In 2011, there were an estimated 10.4 million refugees, 895 thousand asylum seekers, 15.4 million Internally Displaced Persons (IDPs), 3.4 million stateless people and 1.4 million others of concern. IDPs began to overtake refugees as the largest group of forcibly displaced people. The number of people covered by the mandate of UNHCR was growing exponentially.

Refugee Women and Girls in Urban Areas

Around this time it became apparent that many refugees were living as 'urban refugees', usually in slums in host countries. Reacting to the high levels of insecurity and lack of hope in camps, they took their chances in cities (HRW, 2002). Policy frameworks still focused almost exclusively on refugee camps as the normative experiences of refugees and the vast majority of donor funding targeted camp-based populations. There was limited recognition, both within UNHCR and globally, of the particular risks faced by refugees living in urban contexts. We had become aware of conditions in cities while conducting research in Nairobi, Kenya, as part of the Women at Risk (WaR) Research Project in 2003. When commissioned by UNHCR in 2005 to undertake consultations with Chin refugee women living in New Delhi, we had started to explore and challenge the prevailing understandings and discourse related to urban refugees, which was that although difficult and limited, refugees fared better in cities than in camps because of access to education, health services and work opportunities.

At that time the US Committee for Refugees was spearheading a campaign to bring about an end to the confinement of refugees in camps. Known as the End Refugee Warehousing Campaign (Muthee, 2008), it had highlighted the negative impacts of prolonged encampment on refugee communities' rights. Community consultations in urban contexts took place in Mae Sot, Thailand; New Delhi and Mizoram in India; Cairo in Egypt and Kuala Lumpur in Malaysia, as part of our ongoing research with refugee women and girls at risk. The appalling conditions in which people lived in these sites were different, but comparable, to the worst refugee camps. At the request of refugees, we furtively filmed in factories, quarries, farms and markets, where people were forced to work in dangerous conditions for exploitative, or even no, wages. Refugees, both those registered with UNHCR and those unregistered because they could not reach a UNHCR office, discussed the exploitation when work could be found, and lack of access to healthcare and education for themselves and their children. Often the only income for a family was for the women and girls to work as housemaids in unsafe conditions or to sell sex. In particular, the refugees discussed the vulnerability of women and girls, and, in some cases, boys and young men, to sexual abuse and the almost total impunity enjoyed by perpetrators. And this was not recognised or addressed.

Salvation Through 'Candle Making'

This lack of recognition was most strikingly highlighted in UNHCR's Urban Refugee policy, released with much fanfare in September 2009. The policy stressed the importance of active engagement by UNHCR with host country governments, city authorities and, most importantly, with refugee communities, emphasising the importance of community outreach, community mobilisation and regular communication (UNHCR, 2009). In an otherwise rights-based policy, its lack of focus on the particular risks faced by women and girls in urban areas was heartbreaking. It used stigmatising and disempowering language, portraying women who resort to 'survival sex' as 'in need of saving' through the provision of 'constructive and productive alternatives' (UNHCR, 2009, para 81) neither naming survival sex as rape nor recognising that, even in their desperation, refugee women can and do exercise agency. This spawned livelihood projects in many refugee sites, where well-intentioned organisations 'took in' women and girls forced to sell sex in order to provide for their families, and taught them skills such as candle making, as a path to' *reform*'. The return for their labour was so small that they were unable fulfil their responsibilities to parents, siblings and their own children, and they soon returned to selling sex, proving that '*they really are immoral*'.

A significant flaw in the Policy was in the section entitled Age, Gender and Diversity Mainstreaming (AGDM). The only example of heightened risk was for single men in urban settings. While we did not dispute the risks for single men, nor the urgent need to address their protection needs, we were furious that this is the *only* example used. The risks of rape, sexual and gender-based violence faced by women were framed only in association with the likelihood that 'loss of male identity through being forced to undertake low status work may lead to increased risks for women and children of domestic and family violence' (UNHCR, 2009, para. 36).

This framing was deeply problematic, as it placed the lives of refugee women firmly within 'the private sphere' and the agency they exercised for their own and their family's protection was completely silenced. The policy did not recognise the wide range of other contexts in which refugee women and girls faced risks of rape, sexual and gender-based violence in urban areas. Yet this was written in 2009, following a decade of action, outrage and legal advances, including the international definitions of rape and sexual violence in the Statute of the International Criminal Court (ICC) and the adoption of Security Council Resolution 1888, October 2009, which provided an accountability and implementation framework for UNSC Resolution 1820 addressing rape in war (UN, 2009). It highlighted the degree to which international policy-making still remained siloed within different agencies of the UN. These policies were still not consistently recognised within the halls of UNHCR Geneva, let alone in staff practices on the ground. When the lack of gender in the Urban Policy was raised at the ExCom meeting at which it was launched, we were advised by a senior UNHCR staff member that this was no longer an issue as 'gender had been mainstreamed'. 'Yes, mainstreamed into oblivion!' Eileen replied, recalling the words she had first heard in 1989! The High Commissioner's Dialogue (HCD) focused on the implementation of the Urban Refugee Policy. Following intense lobbying, and a very undiplomatic temper tantrum in Geneva, Eileen was invited by the High Commissioner's office to prepare a paper on gender in urban refugee sites and present it in the opening plenary of the HCD.

The paper reflected what the communities had shared over many years. It highlighted that rape, sexual and gender-based violence affected not just the individual woman, but also her family and her community.

The spiralling impact of rape.
Can you imagine how it feels every night when you sit down to eat food with your children and you know that the only reason there is food on the table is that your wife works as a prostitute, or as a housemaid, and every day her employer abuses and rapes her... That is what happens to so many of the

men here, and what can we do? If she does not work, we will not eat – the children will starve ... and if she gets pregnant – who is the father ... is she diseased? These are the things we think the whole time – yet we never talk about it. The shame is too grea.t (Refugee man, Cairo, Egypt).

This was shared in a workshop at a short course at the American University in Cairo organised by Dr Barbara Harrell-Bond. A group of 23 men, refugees or asylum seekers from different African communities, participated. The speaker's voice was breaking with emotion and, when he finished, most of the men were crying and nodding in agreement. We met with them out of hours to discuss their experiences as urban refugees. The issue of rape, sexual and gender-based violence was high on their agenda, and the high levels of 'prostitution' in their communities were quickly identified. Initially, there was a call for the women to be jailed for immoral behaviour. Unpacking and discussing the layers of discrimination the women faced, the men moved from condemning women who worked as prostitutes to devising strategies to assist women who were forced to sell sex. Despite the incredible toll on families and communities, it was the women who continued to bear the brunt of this violence. They contracted sexually transmitted diseases, became pregnant with children of rape or had to undergo dangerous abortions. They were beaten by husbands not able to stand the shame in the eyes of the community and often isolated and silenced, and thus placed in even more danger (Fig. 10.1).

So after being raped, they feel like they are nobody, they have no value, so that they are committing more crimes, they are prostitution, they are risking their lives, because of HIV and violence. She [feels] so much shame ...she might feel like committing suicide. (Chin refugee adolescent girl 2008)

The phrase 'no-one has ever allowed us to talk about this before' was a recurring refrain.

Intersecting Risks

The multiple risks and discriminations faced by urban refugees could not be viewed and addressed in isolation from each other. Again, the theory of intersectionality was invaluable. Access to safe and secure accommodation was identified as critical to the protection of women and girls from rape,

Fig. 10.1 Storyboard—Risks for women, Mizoram 2008

sexual and gender-based violence. It was crucial for the maintenance of good health and to provide an environment allowing children and adults to study. Families struggled to maintain normal familial relationships if they were not afforded some level of privacy. Without safe accommodation, women were more likely to seek work in unsafe places, including in illegal factories, which provided some basic accommodation, but also a high risk of rape. Women and girls who were raped faced many health challenges, but no access to adequate livelihoods or free health services usually meant that they were not able to seek medical attention. Poor health outcomes led to an inability to protect children or to access what little employment there was. And so the cycle of risk and vulnerability continued.

Staff Attitudes

A compounding problem was the impact of the critical and judgmental attitudes of some UNHCR and NGO staff towards the refugee communities. There were certainly many caring staff who regularly went above and beyond the call of duty, finding ways to navigate often complex and bureaucratic systems to provide direct support and protection to refugees. One memorable example came from the UNHCR office in Turkey where, every pay day, the staff would collect money to pay for accommodating unaccompanied minors for whom they could not find any other help. Other staff became part of the problem, as can be seen in these quotes.

> *Why should we speak to you, you are a dirty prostitute.* (UNHCR Implementing Partner 2005)
>
> *Don't lie to me about being raped – that will not get you refugee status – I suppose you just want to be resettled – do you think we would not want to live in the USA too? Why should you go?* (Local staff member of an NGO following rejection of a refugee resettlement application by UNHCR)

While examples such as these were often attributed to local staff, what became increasingly clear was the way in which systemic racism and discrimination within the humanitarian system entrenched these attitudes. There is a hierarchy within international organisations which privileges international staff over local staff in everything from salary to accommodation standards and leave arrangements. On our first visit to Kakuma, we saw 'Soweto', the informal 'slum community' within the NGO compound to accommodate the families of poorly paid local staff. Where international staff went for 'rest and recreation' every eight weeks, returning to Nairobi by plane, local staff had only annual holidays and a dangerous two-day journey back to Nairobi by bus. This left many local staff feeling undervalued, disenfranchised and burnt out. Within both UNHCR and among NGOs, there were staff who displayed outstanding skill and commitment to their work. There were others who used notions of charitable or religious practice as excuses for substandard service provision (Pittaway, 2009).

Developing the Age, Gender and Diversity Matrix Tool

In December, while attending the High Commissioner's Dialogue (HCD), we also attended a workshop on Age, Gender and Diversity convened by the new UNHCR Senior Coordinator on Gender, Luisa Cremonese. One of the workshop aims was to develop a method to measure progress on the implementation of the UNHCR Five Commitments to Refugee Women. She noted the need to link these directly to the AGD mainstreaming approach in order to assess progress in realising the rights of refugee women and girls across the age spectrum, and from diverse groups. Eileen was sitting beside me and started scribbling madly. She passed me her notebook in which she a drawn a matrix listing the indictors Luisa had mentioned, such as access to sanitary materials, leadership and livelihoods, listed down the left axis, and the different AGD groups on the top axis e.g. Girls 0–10, Girls 11–17, women with a disability and LBTI women. This matrix became a key research method. I first used it in early 2010 with a group of young Chin/Burmese refugee women living in New Delhi. Participants were asked to discuss what impact each of the issues listed on the left axis had for the different groups on the top. As the women presented each section of the matrix, it became clear the method would be a powerful new addition to the Reciprocal Research methodology, eliciting the different impacts of protection risks across age and diversity groups.

Refugee Women as Part of the Solution

While we were still in Geneva, a UNHCR staffer working in Syria gave an inspiring presentation, describing a project highlighting the critical role the refugee women employed as outreach workers had played by using their local knowledge of the risks women faced. The presentation was significant because it recognised and supported the skills and agency of refugee women. It had recently ceased due to lack of money, so we undertook to speak to the Australian Government about the possibility of providing some targeted funding. They were willing to consider it, but the closing date for the funding round they nominated, late one evening, was just over 24 hours away! Early next morning, work began on a proposal for a small pilot project for $18,000 directly linking the work in Syria to our previous work on the Heightened Risk Identification Tool (HRIT). Feedback from UNHCR field staff had indicated the tool was too complex. The aim was to work with the team of refugee women outreach workers in Syria to develop a simpler version, to be used by refugee community workers. After reading the application form more closely, it became apparent we had been asked to apply for something much bigger. Time was running out, we but the completed application for almost $400,000 was lodged with only minutes to spare. It was fully funded in June 2010.

Excitement at receiving this funding dissipated as the political situation in Syria quickly started to unravel. The advice was not to travel there. I was in Geneva at the time and suggested to Luisa Cremonese, the Gender advisor, that we meet with Erika Feller, and ask her advice. In yet another example of the 'profane' operation of the policy-making process, Luisa laughed and said, 'Well Linda, you know that you have a much better chance of getting a meeting with her at such short notice than I do'! This proved true. The next morning, Luisa and I were whisked into a quick meeting with Erika, where we explained that the Project could not go ahead in Syria, and Erika facilitated the moving of the project to Jordan.

Survivors, Protectors, Providers: The Regional Dialogues with Refugee Women and Girls

In 2010, plans for the commemoration of the 60th anniversary of the Refugee Convention in 2011, were tabled, and Erika Feller, always a strong supporter, suggested we think about ways of including women and girls in the proceedings. After discussion with Luisa Cremonese, we were back in Geneva in late September negotiating a contract to undertake a series of Dialogues with refugee women in seven countries, to ensure that the protection concerns of women and girls were heard and their recommendations on how to address problems recorded and shared with UNHCR, NGOs and host country governments. There were only seven months to do this before the report would be tabled at the June 2011 Standing Committee! The Dialogues, which each lasted four days, took place in Colombia, Finland, India, Jordan, Thailand, Uganda and Zambia; in three refugee camps; two urban refugee sites; one urban IDP site and one country of resettlement. Colleagues from UNHCR,[i] as well as many research associates and student interns, joined us for this project. Without them, the Dialogues could not have happened.[ii] A further Dialogue was held in Australia for resettled refugee women. This was linked to a very successful international refugee conference, organised and coordinated by our CRR and ANCORW colleagues, Geraldine Doney and Effie Mitchell.

Using the matrix exercise, the Dialogues focused on ten core protection areas proposed by UNHCR: individual documentation; women's representation in leadership; sexual and gender-based violence (SGBV); shelter; sanitary materials; health issues; legal remedies; education; economic self-reliance; and general violence. (Fig. 10.2). The geographical reach of the project provided the opportunity to hear the voices of refugee women and girls from around the globe. In total, over 1,000 refugee, asylum-seeker and internally displaced women and girls, and 200

Fig. 10.2 The matrix tool

men and boys, were consulted on their protection needs over four days of human rights training and situational analysis workshops. A total of 120 UNHCR staff, 52 Government officials and over 90 staff from 55 NGO partner agencies also participated in the Dialogues.

The reports produced from the Dialogues were designed to assist governments to formulate pledges relating to improving the legal, social and physical protection of refugee women and girls and present them at the High-Level Ministerial Meeting to be held in December 2011, as the culmination of the 60th anniversary celebrations. A comprehensive report was prepared, summarising the key findings and recommendations from seven sites and over 98 days of intensive consultations. The report was presented at the June Standing Committee in preparation for the December meeting. Eileen and I had been allocated time to present it in a plenary session. We argued that it should not be us presenting, but women who had participated in the Dialogues. After a long hard fight, with fantastic support from staff within UNHCR, this was finally agreed and a group of ten women travelled to Geneva to speak on their own behalf. Sadly, women from India and Thailand were not allowed by the Governments to travel to Geneva, or rather, to return if they left the country.

The women representatives' visit to Geneva was timed to coincide with two important annual meetings: a Standing Committee meeting of UNHCR's Executive Committee, 21–23 June; and UNHCR's Annual Consultations with NGOs, 28–30 June. Over two weeks, the ten women made multiple presentations about the lives of refugee women and girls and their families, including recommendations to improve protection. They attended informal meetings with senior UNHCR staff to advocate on specific and localised issues. The Dialogues provided an extraordinary opportunity to influence global processes before an international audience. It was also an occasion to evaluate the 2001 Five Commitments to Refugee Women.

Presentation to the Standing Committee by Refugee Women Representatives

In the Standing Committee Plenary session, five women spoke on the ten core protection areas highlighted in the Dialogues. This was a rare occasion on which any person other than UNHCR and government representatives had been invited to address the Standing Committee from the stage. The presentation was very powerful. Looking around the room from the stage, kneeling out of sight to drive the PowerPoint slides while the women presented, Eileen noticed a strange and unusual silence in the chamber, and looking up, saw that the majority of the audience were moved to tears. What a change from the almost emotion-free zone that generally characterised this meeting! More than 20 governments, equally spread between the global south and the global north made interventions. They thanked the women for their courage and made statements of support for refugee women and girls, and commitments to finding solutions to address the problems they faced. A number of states called on all governments, international organisations, regional organisations and the international community to work together with UNHCR to tackle sexual and gender-based violence. The High Commissioner made an immediate pledge of ten million dollars and asked others to do the same. There were expressions of hope that these messages would be reflected in the promises to be made by governments at the UNHCR High Commissioner's Ministerial meeting in December 2011.

Women as Leaders

The ten women representatives were invited to a panel discussion about participation and leadership chaired by Louise Aubin, then UNHCR's Deputy Director of the Division of International Protection. Louise asked questions about the women's experiences of leadership, the barriers they faced in being leaders, and the qualities of women leaders.

> *A woman that is a leader she can't sleep, she can't eat, she can't work without thinking about the others. It is just giving a hug, to share the sorrows, and also the happiness of the person she has beside her ... the one who is really a leader, she will overcome the problems of all her community. But men, they are like a horse with blinkers on. Women they look everywhere. They look right, they look left, they look up, they look down, they look in front, they look behind, and to the sides. They think of everything and of everyone. In this dialogue you can find them, a group of women leaders, beautiful. They know how to carry on their shoulders a very important responsibility. They are a very important example.* (Refugee woman Geneva 2011 in CRR, 2011)

Their responses highlighted the enormous challenges, risks and barriers faced by displaced women, in both formal and informal roles, as leaders in their communities.

A 'Conversation' with Erika Feller

During the meetings, some suggested that the women had been 'scripted' to always discuss rape, sexual and gender-based violence. In a very well-attended, challenging, confronting and completely unscripted 'open conversation', facilitated by Erika Feller, the ten refugee women shared their experiences and the harsh reality of their lives in refugee camps and urban sites. Erika valiantly tried to steer the topic away from rape, sexual and gender violence and focus on other key issues of concern, such as education, shelter and health services. In response to her questions, the sexual abuse the women had suffered in all aspects of their lives was central to their answers. All of them reiterated that they had to be consulted and included in decision-making about their lives and be active partners in solutions to their problems. Erika Feller commented very frankly that *'the Dialogues had been a lesson about our collective failure to address these issues over the past decade'*. (CRR, 2011)

Aftermath Resettlement Meeting

Immediately after the Dialogues, UNHCR held its Annual Tripartite Consultations on Resettlement (ATCR) meeting, chaired, on this occasion, by the USA. Their representative to Geneva invited Eileen to the meeting to make a presentation about the Dialogues and the WaR Program. She talked about women who had been forced to sell sex to survive and to provide food for their families and who were often excluded from the WaR Program because they were seen to be of immoral character. In fact, they were often the most at risk of all women. The representative from the USA was horrified to hear this and confirmed that the USA did prevent women who had been 'prostitutes', as they named them, entering the USA on the grounds that prostitution was criminal behaviour. She undertook to ensure that women who had been forced to sell sex would not be covered by this piece of legislation and would be considered for resettlement as a matter of urgency in future. This was an example of the power of being able to attend meetings and take real-life examples of what is happening on the ground to the decision-makers.

The Major Messages from the Dialogues

Four overarching insights emerged. They were: the commonality of protection challenges in all regions; the intersectionality of the protection challenges; the specific situation of refugees when compared to local populations and the need to improve accountability to women and girls. In one of the Dialogues the women stated, 'We have to talk about everything – one thing leads to another – we have to break the cycles'. The most striking aspect of the outcomes was the similarity of experience reported by the refugee and internally displaced groups across very diverse settings. At one point during the fifth of the seven Dialogues, Eileen commented, 'you know if we just shut our eyes and listen to the interpreters we could be in any of the countries in which we have already conducted Dialogues. The issues are the same in every place'.

Unsurprisingly, the most distressing reports heard in every country were the endemic and serious levels of rape, sexual and gender-based violence experienced by women and girls, boys and young men, and the inadequate prevention or response mechanisms. Participants in every site observed that the capacity of refugees to be involved in their own protection was not generally recognised or utilised, and that the difficult communication with some staff (including UNHCR, implementing partners and host governments) was exacerbating their problems. What was really strong was the women's own analysis of the intersectionality of the ten core issues across all sites, highlighting, yet again, how a failure in one area compounds problems in access to protection in all other areas.

The Final Report, Survivors, Protectors, Providers

The report (UNHCR, 2011) categorised the findings, challenges and recommendations under each of the ten core protection areas set out above. It was signed off by a famous actor, the UNHCR goodwill Ambassador at the time. The solutions suggested by the participants in the Regional Dialogues had the clear potential to address many of the protection concerns discussed. Very little of the evidence we collected was new to us. What was shocking was that it was new to so many people in the protection field, from those on the ground to those at the very top. The Matrix exercise helped to unpack these problems and to deepen understanding of their different manifestations across the age, gender and diverse groups.

Rape, Sexual and Gender-Based Violence

> *We have many live examples about women who have survived these rapes and have brought up children [conceived] from these attacks .... 'We live in constant terror, unable to protect ourselves and our children'.* (UNHCR, 2011)

Women and men in each country reported that rape, sexual and gender-based violence, was endemic and occurred every day and in every aspect of the lives of women and girls and their families. The heightened vulnerability of unaccompanied minors and women with a disability was repeatedly raised. Boys and young men were also targeted for rape, and families simply did not know how to respond to the needs and trauma of abused girls and boys. Humanitarian workers were overwhelmed by the sheer scope of the problem but were not speaking out about it. Credible complaints mechanisms, and services which respond to sexual and gender-based violence in any meaningful way, were scarce, and adequate physical and mental health services and support were not systematically available.

Women were afraid to access these services, especially when staff made them feel ashamed, as if the rape was their fault.

Individual Documentation

> *My sister is resettled – she says that there they have to register dogs or they get into trouble with the police. Here we cannot even register human beings.* (UNHCR, 2011)

The Matrix tool brought many unexpected issues to light. One striking example was the impact of lack of documentation on children zero to five. In the first dialogue in New Delhi, it became apparent that the lack of birth registration in India had devastating consequences for families, who could not register their babies for food, health care and education without a birth certificate.

Discrimination, even in death

Those with babies who died without having their births registered, could not register their deaths, and without a death certificate, were not allowed to bury them in cemeteries. The women recounted heart-breaking stories of having to sneak into public parks at night to bury their dead babies.

Women's Representation in Leadership

> *To come to this training, it costs the same as one egg. I have two teenage sons who work as labourers. Every week I try and buy one egg to share between them on Sunday. So I have to choose between the egg for my sons or to come to the training.* (UNHCR, 2011)

Barriers to women's leadership emerged as a very strong theme across all the Dialogues. Women reported they were often marginalised by men in their homes and routinely excluded from meaningful positions in their own communities. Many were denied education, preventing them learning the language (e.g. English) used in meetings. This lack of leadership opportunities was reinforced by the often 'tokenistic' recognition and inclusion in decision-making processes in the public sphere. Just being appointed to committees was not enough. They needed training to 'give them the right words to use', help them to understand their rights and instil more self-confidence. Furthermore, many women needed childcare to be able to attend meetings, assistance with transport, and compensation if attending meetings caused them to lose scarce income. Meeting places had to be safe for them to visit. They argued that if women do not participate in decision-making, the specific needs of women and girls, in particular the most vulnerable, are often not adequately acknowledged or addressed, with long-lasting consequences. Without this support women's knowledge and expertise is lost.

Shelter

> *I would like to have a house that has a door so I can keep my daughters safe.* (UNHCR, 2011)

In both urban and camp situations, refugees had major problems with overcrowded, unhealthy accommodation, which was often not weatherproof. In many cases, large families were forced to live in small, unventilated spaces, along with their domestic animals. Women with a disability, older women and members of the LGBTIQ+ communities had particular problems finding safe shelter because of the negative attitudes of landlords and other refugees. In both urban and rural sites, access to adequate clean water and sanitation was a major problem. Lack of simple security measures, such as locks for doors, created serious protection problems for women in camps. In urban areas, because of the high costs of rental, families were often forced to share accommodation, with 12 or more people living and sleeping in one cramped room. Unaccompanied minor girls in urban areas were often forced to share rooms with single young men and, in both camps and urban settings, to work as domestic servants for families. The lack of privacy contributed to family breakdown, as married couples had no privacy for intimacy, and women were ashamed to undress in rooms where their children also lived. This posed particular problems for woman and girls during menstruation (Fig. 10.3).

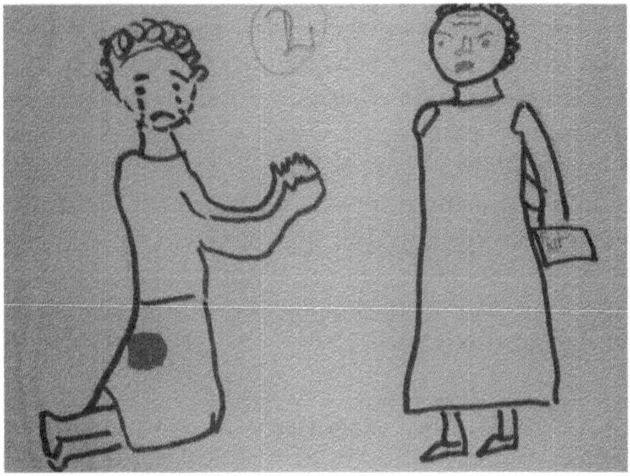

Fig. 10.3 Storyboard—Lack of sanitary materials

Sanitary Materials

> *The sanitary materials are insufficient, and not all women get them. We have no soap and only one bucket for water, for food, for washing ourselves. How can we wash our rags in that bucket? Women are getting infections.* (UNHCR, 2011)

Lack of sanitary materials was a major, perennial problem for the majority of refugee women and girls. In some sites, no sanitary materials were available and in others the quantity and quality were insufficient. Many women supplied with reusable sanitary materials did not have sufficient soap or a bucket in which to wash them. In most sites, only women from 12–49 years of age received sanitary materials and yet some girls of nine menstruate, as do many women over 50. The lack of sanitary materials prevented some girls from attending school and women from leaving the house during menstruation. This problem caused great shame.

Health

She was twelve years old when she got pregnant and she died because her hips were not wide enough to give birth. (UNHCR, 2011)

A key concern for women and girls was the lack of effective reproductive health care. There was concern in all sites about the large number of young girls, raped or in child marriages, who died in childbirth as they were physically too small to bear a child. Women requested that doctors be able to perform Caesareans in such cases. The death of young women in childbirth had come up so often in the Dialogue held in Uganda that we requested access to the health reports on the incidence of adolescent girls dying in labour. It was stated that they did not collect that kind of data. Such lack of data disaggregated by age, gender and diversity had a catastrophic impact on policy and practice.

No-one understands what we need – so they give us Panadol [paracetamol]....'

Medical treatment for women and girls who had suffered rape was often inadequate or unavailable. Racist attitudes from some medical staff, and lack of interpreters for medical cases, seriously impacted on the level of care received. Lack of specialist care for people suffering from trauma and torture and the impacts of rape, sexual and gender-based violence, was discussed in every dialogue. Other issues included a lack of medicines in health centres, with reports of paracetamol being the major drug issued, often inappropriately, for serious diseases, following rape and for mental illness. We could fill a book with quotations about paracetamol being the only medicine on offer, so often did this issue arise in consultations.

Legal Remedies

Even if the woman has the money to take the man to court... the person who is taken to prison will stay for a few days then return to their settlement. (UNHCR, 2011)

Access to all aspects of legal protection was a major concern. Refugees reported racism and corruption in the police forces in several countries. Abuse included beatings and rape. Many could not access or afford lawyers, and access to mobile courts was limited. The delay in response to criminal matters meant that often no action was taken. There was a general distrust of camp justice systems run mainly by male leaders. One woman at the dialogues told of being put in a cell for the weekend after she reported being raped. The police then arrested the perpetrator and put him in the same cell with the woman.

Education

> *When a girl is raped at school, she drops out and refuses to go to school and her mother is afraid to send her other daughters to school.* (UNHCR, 2011)

Refugee communities identified access to education for children and adults, in particular for girl children and women, as a critical part of planning for successful durable solutions. Barriers to achieving this included overcrowding in schools, with classes of up to 300 children reported; inadequate school buildings, sometimes without water and sanitation; unqualified and poorly trained teachers; high levels of violence, including rape and sexual abuse of girls and boys by teaching staff and local pupils, with 'sex for grades' reported to be a common practice. There were few female teachers, and pregnant girls and those with babies were not allowed to attend school. When vocational and adult education was available, it often targeted males or was given at times when women could not attend because of family duties. Little attention was given to the needs of pre-literate women. Former qualifications and levels of education were often not recognised by host countries and language was an obstacle for many students.

Economic Self-Reliance

> *For girls 12 to 17 years old, not all families can support all family members. The girls might be forced into prostitution to support the family. They also risk forced marriage to older husbands or to local men to obtain residency. This is an early age to be responsible for a family.* (UNHCR, 2011)

Lack of access to employment and legal means of income generation was one of the biggest hardships reported by refugee women and girls. Lack of permission to work in many host countries and lack of access to adequate subsistence allowances or rations, left many women in desperate situations, unable to pay rent, purchase food or other essential items. It impacted on health and the ability to send their children to school. When vocational training was provided, it was often not targeted to the local labour markets, and income generation schemes did not allow refugees to generate survival incomes.

General Violence

For us women – all violence is sexual violence, they rape us everywhere. (UNHCR, 2011)

While sexual violence was the key issue for refugee women, they also discussed a range of other forms of violence faced in camps and in urban areas. These included inter-tribal violence and violence from local populations, such as beatings, burning of huts in camps and refugee accommodation in cities. In several camps and one urban area, the issue of witchcraft was raised, with reports of women being targeted for violence if they were named as witches, but, on the whole, the communities were afraid to discuss it. The violence of racism and xenophobia was raised in every site. This affected women and girls when they were targeted for sexual abuse because of their difference from the local populations.

The Value of the Dialogues Report

The Dialogues Report was, in itself, a critical advocacy tool as it provided a comprehensive body of evidence about the common human rights abuses faced by displaced women and girls. It highlighted what little progress had been made in implementing the Five Commitments to Refugee Women and the magical disappearing trick of 'mainstreaming'. On the positive side, it demonstrated that the process was as important as the outcome. Many in the humanitarian field had actually listened to refugee women for the first time. Following the progress of the women as they returned to their lives as refugees, all continued their work as inspirational women leaders. The success of the Dialogues had exceeded expectations. People had listened and there had been an instant response. The challenge now was to see it carried forward into future practice.

Conclusion

After the very successful dialogue in Jordan (replacing Syria), the Government of Jordan changed its policy on research by humanitarian agencies and academics. We were collateral damage. Our Project seemed doomed.

Notes

1. UNHCR Geneva staff: Luisa Cremonese, Astrid Van Genderen Stort, Irene Van Horssen, Greg Garras, Chiara Cardoletti, Jose Riera, Ita Sheehy and Stefanie Krause and the UNHCR Country Representatives and their teams in each site.
2. Geraldine Doney, Effie Mitchell, May Maloney, Rebecca Eckert, Marcela Garrett, Kristy Ward, Meg Davis, Aileen Crowe, Carmel Clarke, Natasha Schofield, Lany Arévalo, Kate Clarke, Melika Sheikh-Eldin.

References

Ferris, E. (2014). *Ten years after humanitarian reform: How have IDPs fared?* The Brookings Institution. Retrieved November 15, 2021, from https://brookings.edu/wp-content/uploads/2016/06/Introduction-to-final-report-IDP-Study-FINAL.pdf

Human Rights Watch (HRW). (2002). *Hidden in plain view: Refugees living without protection in Nairobi and Kampala*. Human Rights Watch. Retrieved November 12, 2021, from https://www.hrw.org/reports/2002/kenyugan/

Muthee, A. W. (2008). *End refugee warehousing*. Capstone Collection 2118. Retrieved November 15, 2021, from https://digitalcollections.sit.edu/capstones/2118

Okoth-Obbo, G. (2009). *Introduction by George Okoth-Obbo, Director, Division of International Protection Services. Report on Age, Gender and Diversity Mainstreaming and Protection of People with Disability in the oral update on "Global Programmes and Partnerships – Strategic Presentation Summary", 44th Meeting of the Standing Committee, 3–5 March, 2009*. UNHCR. Retrieved October 25, 2009, from http://www.unhcr.org/refworld/pdfid/4a5c51d12.pdf

Pittaway, E., & Bartolomei, L. (2003, November). When death is the only medicine: Health risks on the Thai Burma border. *The Health Exchange*, pp. 10–13.

Pittaway, E., & Bartolomei, L. (2009). Innovations in research with refugee communities. *Transitions*, 21, NSW Service for the Treatment and Rehabilitation of Torture and Trauma Survivors (STARTTS). Retrieved December 20, 2010 from http://www.startts.org.au/default.aspx?id=405

Pittaway, E. (2009). *Making mainstreaming a reality—Gender and the UNHCR policy an refugee protection and solutions in urban areas.* UNHCR. Retrieved December 20, 2011, from www.unhcr.org/4b0bb83f9.PDF

Pittaway, E., & Bartolomei, L. (2005a). *Identification and protection of 'women at risk'—Summary report.* Centre For Refugee Research, University of New South Wales.

Pittaway, E., & Bartolomei, L. (2005b). *New Delhi: Report to United Nations High Commissioner for Refugees.* Centre for Refugee Research, University of New South Wales (On file with the authors).

Pittaway, E., & Bartolomei, L. (2005c). *Deidentified report to United Nations High Commissioner for Refugees.* Centre for Refugee Research, University of New South Wales (On file with the authors).

Pittaway, E., Bartolomei, L., Younes, R., & Lin, J. (2003). *Seeking hope, seeking safety: Refugee women at risk.* ANCORW & CRR.

United Nations. (2009). *Resolution 1888.* Adopted by the Security Council at its 6,195th meeting on 30 September 2009. Retrieved November 24, 2021 from http://unscr.com/en/resolutions/doc/1888

United Nations High Commissioner for Refugees. (2009). *UNHCR policy on refugee protection and solutions in urban areas.* UNHCR. Retrieved November 24, 2021 from https://www.unhcr.org/protection/hcdialogue%20/4ab356ab6/unhcr-policy-refugee-protection-solutions-urban-areas.html

United Nations High Commissioner for Refugees. (2011). *Survivors, protectors, providers: Refugee women speak out.* UNHCR. Retrieved November 24, 2021 from https://www.unhcr.org/50f919f39.pdf

CHAPTER 11

Anger and Despair—Naming the 'Elephants in the Room'—2012–2016

Narrated by Eileen Pittaway

A Chapter of Reflection... and a Miracle!

After the euphoria of the Dialogues, where the women had spoken so eloquently and the world had cried with them, very little actually happened. Despite the best intentions of so many in UNHCR and the humanitarian aid sector, many things were not changing, and the amazing work done by so many refugees was all too often not acknowledged. There were so many barriers, and an apparent acceptance that they were just too hard to address. All, including the refugee women, swung between anger and despair at the lack of progress.

An Increasingly Unstable World

As we started our journey, there were an estimated 8 million refugees, with 3 million people internally displaced. The world was a hopeful place. There was talk of world peace, stability and an end to absolute poverty. However, since the turn of the millennium, that hope had faded. The power balance in the world was changing and inequality was rising in both

The original version of this chapter was revised: Author name has been updated in this chapter. The correction to this chapter is available at https://doi.org/10.1007/978-981-19-0916-0_14

E. Pittaway and L. A. Bartolomei,
Only Rape! Human Rights and Gender Equality for Refugee Women,
Sustainable Development Goals Series,
https://doi.org/10.1007/978-981-19-0916-0_11

the global north and the global south. The impacts of the 2008 Global Financial Crisis were still felt, resulting in cuts to foreign and humanitarian aid. Global terrorism, a rise in nationalism, and the climate crisis combined to create an air of pessimism and despair.

In the late 1980s, there had been hoped that refugee flows would diminish and solutions would be found for the remaining refugees still in asylum countries. By 2016, this had proved to be a fool's dream. The years between 2012 and 2016 were tumultuous, with an increase in conflicts and displacement, and a decline in political will to receive refugees. There were multiple civil wars in African nations, with millions fleeing to adjoining countries to seek refuge. Many African countries both produced and hosted refugees. Sudan gained independence in 2011, but violent conflict again erupted in 2014. At the same time, the Western African Ebola virus epidemic was causing major loss of life and socioeconomic disruption in the region. Conflict continued between Israel and Palestine. In 2015, war broke out in Yemen, the 'Arab Spring' had destabilised swathes of the Middle East and North Africa, bringing civil wars and brutal reprisals from authoritarian regimes, which swiftly put down the nascent pro-democracy movements. The exodus from Syria had begun and would lead to 5.6 million people fleeing the country, with millions more displaced inside. Many people made horrendous journeys on foot, with small children, over mountains and rough terrain. Others took enormous risks to cross the Mediterranean in small boats. Huge numbers sought refuge in Europe, causing a major political pushback (Table 11.1).

Increasing need, decreasing funding and burgeoning right-wing politics made advocacy harder than ever. This greatly impeded the response by UNHCR and the humanitarian aid sector to the huge increase in forcibly displaced peoples. It threatened to overwhelm the needs of refugee women and girls: programs to respond to rape, sexual and gender-based violence, were still chronically underfunded, with less than 1% of global humanitarian aid funding being spent on prevention and response. This would be the case for some years to come (OCHA, 2019). Significantly,

Table 11.1 Refugees and forcibly displaced peoples under UNHCR's Mandate, 2011–2016(UNHCR, 2021a, 2021b)

Year	*Refugees and forcibly displaced peoples*
2012	46 million
2013	52 million
2014	60 million
2015	65 million
2016	65.5 million

by 2016, the proportion of refugee women and girls enduring rape and sexual violence was as high as it had been at the beginning of our story. It was telling to note the difference in attention and the response given to refugee movements which directly impacted on the developed nations of Europe and the lack of attention to the millions of people displaced in economically depressed African nations at that time.

The Impact on the Refugee Sector

In humanitarian aid discourse, the phrase 'women and girls' became generalised to 'gender', not because of a strong feminist analysis of gender identity, but because the backlash against the focus on women and girls was so great. Worst of all was the still pervasive view that *'gender is mainstreamed now!'* when it patently wasn't! Comments in meetings such as '*Oh, Linda and Eileen are here—we better make sure women are on the agenda*', said with a smirk, emphasised the attitude of many in the sector. We would fight to get rape, sexual and gender-based violence, onto the agendas of international meetings, only to find that, at best, it was a side event. Time and time again, we sat in sessions where the issue was raised and discussed (a minor triumph!), but when the meeting reports came out, not only were they missing references to women and girls, but they were also silent on the discussions that had taken place. This was not universal. Some member states, INGOs and local groups were putting themselves on the line to support and protect refugee women and girls, but they were struggling to be heard.

Refugee communities were losing hope. Many people no longer believed the stories of better futures and the power of human rights. Academics and service providers working with the communities were at best tolerated, and at worst viewed with disdain, distrust and even fear. Our shared anger and frustration surfaced frequently during this period. As part of the *Accidental Counselling Training* we provide, we have always taught that it is not good to bottle up anger, so looking back at the major obstacles that refugee women faced and continue to face, we knew some steam needed to be released.

Addressing the 'Elephants' in the Room

I have always liked this metaphor, referring to obstacles that are known, but are never acknowledged. In my 'mind's eye' I see them, large, grey and difficult to move, eclipsing everything else, but never mentioned or addressed because of … *why*? Is it because they are so powerful, so much bigger than everything else, so dangerous, that risking annoying them can cause even more damage than pretending they are not there? Let us imagine that this metaphoric room is *'refugee protection'*. We talk about *the* elephant, yet in this room there is a whole *herd* of them with refugee women and girls struggling to be heard over the stampede! Most are elephants you have already met, but many in the sector refuse to acknowledge them. In order to move forward, they must all be named and confronted.

The Patriarch: UNHCR

The biggest elephant of all is UNHCR itself, or rather, the constant suggestion that UNHCR is not adequately protecting refugees. UNHCR's mandate is determined by its member states, and the diplomatic tightrope it has to walk is perilously thin and long. It is *member states* that first generate refugees, and it is the response of other member states that sometimes fails refugees. Underfunded and understaffed, UNHCR is certainly not perfect, and, at times, it does indeed fail, particularly in its inconsistent response to women and girls. The fact that there has not been a senior-level staff member with dedicated responsibility for refugee women and age, gender and diversity issues since 2003 is the clearest indication of this. Nonetheless, it is important to recognise that UNHCR staff are faced with impossible and horrifying tasks and receive very little support. They work long stints in places deemed too dangerous for family to join them and suffer the emotional toll of hearing so many repeated stories of torture and trauma. UNHCR colleagues have told us of having to go out to pick up bodies in a truck, and seeing colleagues kidnapped, held to ransom and being killed. They do amazing work, get little thanks and a lot of criticism. These things are not addressed, but should be.

The Matriarch: The Humanitarian Aid Sector

Now let us meet the apparently benign 'old woman' elephant, the humanitarian NGO sector. It is huge, almost immovable and certainly resistant to change. How does one criticise the 'mother' of refugee protection? Without these colossal international non-government organisations (INGOs), UNHCR would not function. INGOs run the camps for UNHCR, and, as we have documented in various sites, at times their staff abuse this position of power. In order to obtain the much-needed funding to feed and house the growing number of refugees, UNHCR and the humanitarian aid sector have to appease donors, states and host countries. They cannot afford to annoy some refugee-hosting countries from the global south by talking of rape and sexual abuse, or of the rights of LGBTIQ+ communities, because *it offends their culture*. Similarly, they dare not stand up to conservative leaders in the global north, who, influenced by the Vatican and other powerful Christian groups, repeatedly cut foreign aid for sexual and reproductive health services.

As a result of the neoliberal agenda to reduce state function and to contract out services, the scale and power of INGOs are growing exponentially. Yet too many of them are still—to put it politely—conservative in their practice, and use a 'tick-a-box' approach to gender programming and human rights. Human Rights approach? *Tick!* Gender sensitive? *Tick!*

Tick-a-box for gender and human rights

A Master of Development Studies student, employed by a large international aid agency, was very defensive when we discussed the absence of gender in many humanitarian aid projects. He insisted that gender was always considered in program design, and women's needs were specifically addressed. He did some research and came back shamefacedly to admit that the projects he had examined seemed to embody a gender approach on paper only. He described a well-funded project to bring technology-based livelihoods to a large village in Kenya, training the men and establishing a small factory. In the section of the funding application asking if it was gender sensitive, they proudly answered *'yes, every woman in the village will be given two laying chickens so that she can provide more nutritious food for the family and sell a few eggs each week for income'*. Following this up, he found that the majority of the chickens had been killed to feed the men, and the women had complained that when they made a little money, the men would take it for alcohol. In another example, in a village where the women already raised chickens for the local market, making sufficient money to maintain a degree of independence and power over their lives, the men were funded to build chicken sheds and turn the women's cottage industry into the men's 'viable business', depriving the women of their income. This student became our biggest convert to gender equality in aid provision!

A Seemingly Benign Bull Elephant: Faith-Based Organisations

Faith-based organisations have long dominated the humanitarian and development sectors. There is an assumption that faith-based organisations must be a force for good, but they differ as much as the populations they serve. Some of the best work we have seen is done by faith-based organisations, using their faith and dedication to humanity, interpreting doctrine to address needs without thought of their personal safety. Conversely, some faith-based organisations force people to convert, to pray, before being given aid, and deny aid to raped women and single mothers on 'moral' grounds. These organisations are unable to embrace the notion of human rights, because they believe it clashes with their own moral code, ordained by God. In a human rights workshop in Australia, a cleric once asked us, *'If we believe in human rights, what place is there for God'*? The patriarchal belief systems of religions are also problematic; all major religions are dominated by men, and women are usually excluded from taking senior leadership roles. If your religious structure reinforces gender inequality, how will you address it in your services to refugees? If your religion does not embrace gender diversity, women's right to sexual and reproductive health services, and freedom of choice with regard to abortion, how will you incorporate these? The answer is you won't and, as we have detailed the tactics of the religious right at B + 5, they will subvert the efforts of others who try to do so. The most powerful example of this is the USA's 'global gag' banning funding of projects providing sexual and reproductive rights in refugee and development aid provision as one condition for their support. It was imposed by the Bush administration, revoked by the Obama administration and reimposed by the Trump administration because of pressure from the Christian Right. This directly challenges the foundations of a rights-based approach to refugee protection.

> An intervention made at a UNHCR meeting in 2018, drafted on behalf of members of the International Council of Voluntary Agencies (ICVA), stated '*Moreover, a number of NGOs – **not all** – consider that basic services for refugees should provide access to sexual and reproductive health*'.

The Angry Adolescent Bull: The Rise of Extremisms

Pushing behind the older elephants as its power and confidence grows is the adolescent bull elephant, the rise and power of ideologically driven 'extremist organisations', some at the extreme right, others nearer to anarchy. Many are linked to the US Christian Right whose racist, xenophobic, homophobic and misogynistic attitudes are proudly proclaimed on social media and at rallies. White extremism has also emerged as a force to be reckoned with across Europe, and again we have to ask why? That would be a book in itself, but there is a strong argument that its roots lie in poverty, marginalisation and a feeling of being disenfranchised by the powerful in one's own country—of not fully enjoying the benefits of citizenship one would expect to enjoy. At the same time, globalisation has led to a never before seen mass movement of people across borders. The influx of diverse groups is seen as a threat by those less privileged in their own societies—an easy target to blame for the lack of employment and other social problems. This toxic mix has led to an aggressive response to refugees, who are demonised and denied their rights to asylum. The victimisation and marginalisation felt as a result of languishing in refugee camps and urban slums, denied entry to wealthy developed nations, in turn creates a bubbling cauldron for the radicalisation of refugee youth and their conscription to religious extremism.

The Twin Elephants of Fundamentalist Islam and Extremist Christianity in Refugee Populations

These elephants are not often discussed. They are considered shameful, and the fear is, if they are acknowledged, aid might be withdrawn.

Mothers have asked us how to protect their adolescent sons and daughters. They see how anger, marginalisation, poverty and denial of human rights can lead to radical reactions. In Sri Lanka, in 2000, Muslim Tamil women asked, '*How often can our teenage boys watch their mother and, sisters raped and abused, and babies die from lack of food, and not respond*'? They were terrified their boys would be recruited to join extremist groups—and who knows this better than radical leaders, who target, groom and even kidnap young refugees to join them (InfoMigrants, 2017). While there is a media focus on fundamentalist Islam, right-wing Christianity also poses problems, as refugees are forced to convert to obtain food and aid, and there is conflict between opposing Christian groups (Service provider, Kenya 2003).

In the decade 1970–1980, there was much theorising about power inversion (Janeway, 1980; Willis, 1978), the idea that there comes a point when people are so marginalised and disenfranchised by the dominant social structure, that they create a new one from the bottom up. If you cannot succeed in the accepted hierarchy, turn it on its head and build an inverse power system in which you can be king. When your following gains momentum, you feel strong enough to challenge the status quo. This could help to explain what can happen in some refugee populations (UNODC, 2017). They see no hope of ever improving their status in the current social order, so turn instead to alternatives.

The Most Repugnant Bull Elephant in the Room

The most repugnant elephant in the room is the sexual exploitation and abuse of refugees by those charged with protecting them. How do we reconcile the roles of those humanitarian aid workers, faith-based organisations, UN agencies, police, security forces and peacekeepers, in the sexual abuse of children? Why do we allow organisations to let many perpetrators off with impunity, transferring them to new sites rather than prosecuting them? (Ferris, 2007; Freedman, 2015; UNHCR, 2002). It must stop! How can refugees ever trust us and work with us when this is their experience?

The Young Female 'Refugee-Led' Elephant

A small elephant is the assumption embedded in the discourse of the humanitarian aid sector that 'refugee-led' organisations are always 'good'. The same principle of difference and diversity goes for refugee communities as it does for all other communities. There are superb refugee-led organisations, and there are also dysfunctional and divisive ones. An interesting question is who decides which are 'good' and which are 'bad', refugees or us? What standards do we use to measure them? The answer we propose is the human rights framework. Do they adhere to and apply its universal and indivisible principles? Are they genuinely inclusive? Do they use a genuine rights-based approach to practice? This is neither more nor less than should be asked of all humanitarian actors.

A Quiet Young Female Elephant and Her Mother

One elephant is quieter than the others, not taking up as much space. She is the elephant of rape, sexual and gender-based violence, and she stands beside her mother, the elephant of gender inequality. They are often ignored even by the other elephants. If recognised and allowed out of the room they have the potential to contribute to the positive change we are fighting for. They sometimes see the door slightly ajar and notice increased opportunities for education, and talk of participation, but then the door is forced shut again and they are again an invisible, but uncomfortable presence.

And Last—The Elephant of Denial

Me, an elephant? I may be large and grey with a trunk, but really! The biggest barrier of all is that too many key stakeholders do not want to give up the power they hold as service providers. They want to keep the barriers in place. We have been told many times that refugees are too personally involved, too traumatised, not qualified or experienced enough to provide services; that these have to be provided for them by educated professionals.

> In the 2004 tsunami, the fishing boats of Sri Lankan villagers were left battered and broken on the shores. I will never forget driving past groups of traumatised fishermen, with incredulous faces, crouched on the sand behind ropes, watching white volunteer humanitarian aid workers repairing the boats the men had built themselves, and not being allowed to work with them.

The humanitarian aid sector needs to accept their complicity in oppressing those they seek to help, and instead use their expertise to upskill and empower refugees. The power of this herd of elephants is so great it can cause even those who privately acknowledge what is happening to stay silent when they should speak out.

Giving Up or Moving On?

Well, we have got that out of our system! The evidence presented by the refugee women in the Dialogues had shown, beyond any shadow of a doubt, that rape, sexual and gender-based violence, was the most significant barrier for refugee women and girls. They very clearly stated they wanted, and needed, to be part of the solution. They were equally clear that participation did not just mean advocating at UN meetings. They meant on the ground and in every part of their lives.

Liberation through empowerment

I was alone in the forest, collecting wood, when a young soldier came down the path. I was 46 years old and had been gang raped several times before. My heart sank; I would rather die. He undid his trousers, and, pointing the gun at me, told me to lie down. Suddenly all of the training I had been having about women's rights and empowerment came into my head. I had begun to believe I was not powerless. I stood straight and looked him in the eye. I said '*Go ahead and kill me but you will not rape me while I am alive. Look at you! You could be my son; I have grandsons nearly your age. Would you rape your own mother, your grandmother, what would you do to someone who did that*'? I stood waiting for the bullet but it did not come. Tears rolled down his face. '*They tell us we have to do it*', he said. '*Well, you don't*', I replied. I felt so powerful. I went back and told the NGOs, 'I want to start a group and tell other women how to be strong'. A group met each week, helping each other to feel safe and powerful and to protect ourselves and other women. Now I am a community leader. No one will ever rape me again! (Recounted by a refugee woman from Sudan, resettled to Australia).

From Famine to Feast: New Projects in Australia, New Delhi and Nairobi

In the Dialogues, we had been asked to document good practices, which proved challenging. While we knew from our visits to the field that there were some very innovative and effective examples of programs addressing rape, sexual and gender-based violence, and encouraging women's participation, they were seldom shared. We really wanted to work in partnership with refugees, to document what was working and develop new models based on the solutions suggested by the women themselves.

We requested the Australian Government to transfer the funding from the cancelled project in Jordan to New Delhi. Fortunately, they said yes. With the excellent support of Montserrat Feixas Vihe, UNHCR Chief of Mission and her team, we worked with the refugee women who had taken part in the Dialogue held in late 2010. The aim was to develop and evaluate small refugee-led projects which addressed rape, sexual and gender-based violence and community development activities decided by the women. Linda commented that the work in New Delhi provided the positive emotional boost that she so badly needed, because we were finally working on solutions.

As the project got underway, Mohamed Baaruud, who, with his wife Deeqo, had started a small NGO with the Somali community in Sydney, requested us to partner with them in Kenya on a similar program. Mohamed worked at the Service for the Treatment and Rehabilitation of Torture and Trauma Survivors (STARTTS), one of our important Sydney partners and they joined us in the project. The Australian Government also funded this work in Kenya. These projects enabled our first concerted attempt to test our hypothesis about the skills and capacity of refugee women, and to provide support to refugee-led projects.

Around the same time, we were awarded Australian Research Council (ARC) funding to explore the troubled relationship many refugee communities were having with the concept of human rights, as this was having a very detrimental impact on the settlement of newly arrived refugees, in particular Women at Risk (WaR). As the saying goes, it never rains, but it pours! The new projects ran alongside other local initiatives in which we were involved, and our attendance at UNHCR meetings in Geneva continued. We were running on overdrive.

New Delhi and Nairobi

The Refugee Community Development Project, New Delhi, India

The women in New Delhi were particularly keen for us to return and work with them. Still, when we arrived back one year after the Dialogue, our appearance was greeted with genuine surprise by the Somali and Afghan women, who could not quite believe that we, unlike so many previous research consultants, had actually come back! Some cried when they saw us. In consultation with refugee women and men from the Afghan and Somali communities, the Refugee Community Development Project (RCDP) was quickly established. As none of the refugee community groups could receive funding, it was supported by INGO partner Don Bosco. It was also recognised and promoted by UNHCR.

Effie Mitchell and Kristy Ward joined the team as Project Managers and worked tirelessly to ensure that the refugees received the support they needed. UNSW social work colleague Kerrie James also provided important counselling training and support. Over four years, it involved almost 2500 refugees, predominately women and children, in project activities managed by a team of 17 refugee community workers (Bartolomei & Ward, 2016).

Project Respect, Nairobi, Kenya

The initial aim of this project was to work with Somali refugees living in Dadaab Camp, Kenya, but at the last moment, the Kenyan Government withdrew permission for the project to take place and closed down a number of local projects, as it was suspected they were linked to terrorists. At the request of UNHCR Kenya, the project transferred to Eastleigh, a notorious Nairobi slum with a majority Somali community, including an estimated 250,000 registered and unregistered refugees. Eastleigh had a long history of violence, including rape, sexual and gender-based violence, and discrimination against refugees. The project was auspiced by, and housed in, a safe space in a clinic run by two ex-pat Somali doctors, Dr Abdulkadir Hussein and Dr Maimuna Mohamed Isse, a husband and wife team running a basic medical service and a number of psychosocial programs. The clinic was legally registered, and the doctors were able to receive project funding into their bank account. Marcela Garrett joined the UNSW team and travelled to Nairobi to assist in the setup, with training and ongoing support provided via Skype[1] (Garrett & Pittaway, 2014).

Similar, But Different

Men were included in the projects, both to encourage the participation of supportive male community members and to counter any potential backlash from some in the community should the projects solely focus on refugee women and girls. Using the Reciprocal Research methodology, a rights-based, participatory community development approach was used in both Projects. Support from UNHCR and local partners differed in each site, but was essential to the project's success. Staff travelled from Australia as often as funding allowed and communicated via Skype. In both sites the primary focus was on refugee-led projects, with local support. At first, the refugees found it hard to trust and believe that they really would be equal partners in the projects. Some refugees in New Delhi were initially angry that the Australian Government was funding a project in India rather than increasing the number of resettlement places available. They were desperate, and frustrated. Given the exponential increase in the number of refugees across the world, the competition for the few resettlement places available had increased and waiting times, including for WaR, were longer than ever. Their anger was understandable, but as trust built, it began to dissipate, as a growing number of refugee women and men recognised the opportunities the project offered for livelihoods and education for themselves and their children.

The power of democratic process

In New Delhi, we thought it would be useful to establish a management committee with members from each of the communities represented. When first raised, the refugees asked whom *we* would appoint to the committee. They were stunned when told that they were going to elect the committee themselves. Some objections were raised. '*We don't know the process.' 'Some of us are illiterate'. 'The men won't let the women vote!' 'We are several communities, we don't know each other!'* Together we worked out a process. Two people could be nominated from each community, one man and one woman. People nominated for a place on the committee, sponsored by three others from their community. A meeting was called, and they were given five minutes each to present a case to the whole group. Over 100 people turned up and, after the presentations, we had afternoon tea to give people a chance to process this. Then a formal ballot was held. It went extremely well, and a committee was appointed. At the end of the day, a dignified older refugee from Afghanistan approached us. He held out his hand and, with tears running down his face, said. '*I am 45 years old, and in my country, I was a professional man, but that is the first time in my life I have voted! Thank you'*. This is not to suggest the project was a fairy tale—the committee had its problems, like any committee, but these were overcome, and the project was guided by refugees from its inception and for the following four years.

The refugee communities designed, implemented and managed the programs and identified the skills training they would need for each of the programs. A key feature of the projects in both sites was the women's groups, which provided social and emotional support and built on community links for increased safety and protection. The greatest success was a series of outdoor picnics for women and children. Only boys under 12 years of age were allowed to attend: this rule was to ensure that the more conservative and patriarchal refugee men would allow their wives and daughters to participate. The refugee women leaders employed in the project spent many days going from house to house encouraging women to participate, and placating sometimes protective, and other times controlling, husbands. They attracted 100–150 people to each event, providing the women with a social network, and allowing them to build their confidence and talk in a safe space. As their trust grew, they felt that they could discuss issues about which they had previously remained silent, such as survival sex. Women leaders were invited to attend NGO network interagency meetings, including those convened by UNHCR. This gave them a forum for raising awareness of the levels of rape, sexual and gender-based violence, and other problems experienced by the refugee women and girls, and services listened and responded. In New Delhi, outreach support was provided to women who needed emergency assistance and counselling but could not travel to the RCDP office (Bartolomei et al., 2016).

Different models were used to run peer support groups for survivors of rape, sexual and gender-based violence. In Eastleigh, Nairobi, 42 women were trained to run groups across the sprawling suburb, providing information and support and identifying strategies to address the endemic rape, sexual and gender-based violence, which, at times made it impossible for women to attend training or work.

The communities noted a strong correlation between livelihoods and women's safety. In New Delhi, several livelihood initiatives were implemented. Most successful were the Afghan women's snack shop and embroidery project. Staffed by two Afghan widows, the snack shop provided safe work and essential income. It was located in a busy market area where a large number of Afghan families resided, and near a refugee service. The Women's Social Support project provided cooking classes for women, enabling some women to earn a little income through selling food to other families during Afghan festivals and holidays. Sadly, the political climate for refugees in New Delhi changed considerably in early 2015, leading to harassment of the staff by local police and a decision to close the shop for safety reasons. This demonstrated that livelihood activities in situations of protracted displacement are very challenging and require more than a solid business plan and committed staff. They also need formal work rights, and an enabling regulatory environment (Asylum Access, 2014; Bartolomei & Ward, 2016).

A successful livelihoods initiative
In late 2012, RCDP was approached by an Australian woman, living in New Delhi, with connections to Indian fashion designers, who were seeking skilled women to embroider clothes for sale in top-end retail outlets, with markets in India, Europe and Australia. The Embroidery Project provided 70 women with home-based work at a fair salary. Many Afghan women are highly skilled in embroidery, several international designers hired them to produce crochet and needlework for their designer brands. An internationally renowned crochet artist hired Afghan women involved in the project to crochet flowers for an installation as part of a major art exhibition in New Delhi in 2015.

Both communities identified *education* as a powerful protection tool,

> *... If you really want to help someone, you have to educate them, not just give them money... through education a person escapes the prison they are in.* (Somali refugee woman, Nairobi)

The refugees designed and ran programs to meet community needs, drawing on skills already existing in the community, or sourcing trainers with the help of the NGO partners. Acquiring local language skills was high on the list, as the ability to communicate contributed to women's safety. It increased autonomy and decreased their dependence on husbands and adult sons. Women described how they felt more confident to shop alone at the market, or to make phone calls. For several women, their new literacy skills enabled them to find paid work.

> *Because I was illiterate, I was too scared to go out. There were all these lanes and I could not read signs so I would be scared, I would get lost. Because of the best teachers at RCDP, I can now speak English and read Dari and now I won't get lost.* (Afghan woman, New Delhi)

Groups for men included football clubs, training in the provision of psychosocial support, and, at their request, parenting skills. The men's groups were well attended, and participants began to discuss the importance of preventing rape, sexual and gender-based violence, of education, and of including women in a positive way.

What is a family really like?
How do we know how to parent? Our whole lives we have been on the run, hiding, fighting. Most of us young ones have never seen 'home', we have never lived a normal life. Our fathers were in the militia, we have seen our mothers and sisters raped. What does family mean? We don't know how to treat our wives. We want more for our children. We love them and want to learn how to be better fathers and give our children a better life. (Somali Refugee Man, Eastleigh, Nairobi)

Positive Outcomes from the Projects

RCDP was fully, 100 per cent owned by the refugees and it was done in a very professional way, not like [just] a small social project. It proved that they could do bigger things if given the opportunity.... The right kind of training went into it and support was provided. I also saw that a lot of the individuals who were running the project, did it in the most professional way, like any other organisation would do it. (NGO Partner New Delhi)

The evaluations of the projects were very positive. Participants felt an increase in their confidence to confront their problems and to negotiate with service providers. They felt that they had more knowledge and information about the world in which they were living. They reported that the various activities provided social and emotional support as well as building skills and leadership potential. Respect and trust increased between the refugees and the service providers involved in the project, and refugees felt more able to request assistance when needed. For those employed in the projects, the skills they learned provided pathways to paid employment. A number of the men in the project learned more about the meaning of gender inequality and several became strong gender 'champions'. Peer support for survivors of rape, sexual and gender-based violence, made the women feel stronger and more secure.

Women taking on the patriarchy
A group of very brave women working to prevent the sexual abuse of young children in 'madrassa' boarding schools in Nairobi received funding

as part of the project. They were dismayed by reports of sexual assault of children by the male teachers in these schools. The women came together, with the help of doctors Hussein and Isse, to be trained to teach the Koran. They then opened a madrassa boarding school themselves. To do this, they had to challenge the patriarchal structures and the power of the religious leaders. Fortunately, some influential men in the community were supportive and the project went forward, despite objections from other groups of men. The women were determined to challenge what they saw as criminal behaviour, totally divorced from the religion and culture they loved.

The Key Learnings from the Projects

Evidence from the two case study sites in Delhi and Nairobi highlighted the positive impacts that genuine rights-based community development can have on fostering autonomy and self-determination, and in building resilience, positive identities and feelings of self-worth. It provided refugees some control over their lives in protracted exile. One key learning was the capacity of refugees to undertake this work with grace and good humour. It was not all hard labour, and much laughter was shared. The involvement of some key males in the refugee community opened many doors and the men also benefited from involvement in the activities, demonstrating the importance of inclusive community development.

The refugee women and men employed to work in the projects needed more support than originally estimated because of the massive structural barriers to programs, such as racism, lack of work rights and the inability to open a bank account. They had to navigate many challenges in their dual roles as both refugees and refugee community workers. At times, their community members placed demands on them they were unable to fulfil. At other times, the problems they were hearing paralleled their own and they sometimes needed psychosocial support. Success was dependent on the commitment of a large group of refugee women who assumed the mantle of trainers and community leaders with very little remuneration. This caused resentment, as they were doing similar work to that of the local NGOs. The fact that skilled refugees and refugees employed in projects were often the first selected for resettlement, decreasing the social capital in communities, has to be factored into program planning.

Another key lesson was the importance of partnerships. As refugees are unable to register organisations, the willingness of the NGO partners in Nairobi and New Delhi to participate in the project and receive the funds was essential. The support of UNHCR and NGOs in both sites gave the projects a credibility and acceptance in the 'humanitarian space'. The benefits were mutual because the project provided training and some additional income for their staff and volunteers. Involvement in the projects also provided key stakeholders the opportunity to revisit their own models of service delivery, highlighting gaps and problems. Their feedback was uniformly positive.

A further big lesson was the importance of trust and reciprocity. We experienced the damage caused by researchers and NGOs flocking to visit refugee sites and interview refugees, often making promises and never returning, or offering only the hope that by sharing their time and stories, services for other refugees would be improved in the distant future. The lack of continuity in NGO staff and service provision also contributed to the lack of trust. In so many humanitarian aid projects, especially those staffed by expatriates, there is a high turnover of staff. New workers do not necessarily share the same ideological beliefs as previous personnel, and often function as 'street level bureaucrats', interpreting and enacting policy through their own individual lens. Programs are changed or just disappear.

Finally, training and skills development has also resulted in positive impacts for former RCDP Community Workers resettled or voluntarily repatriated. Several are now employed in the USA and Somalia as NGO coordinators, government workers and health staff. These positive impacts highlight the long-term benefit of skills building in an environment where refugees are able to lead and manage projects independently. They show that investment in countries of first asylum through community development approaches can have positive benefits for those able to access durable solutions.

Hakima: the power of a woman!

I was 18 when I fled the war in Somalia. Every night, homes were invaded and we heard the terrible screams of women and girls being raped. Their husbands and fathers just had to watch because they had no weapons to stop it. It was awful, but during the day nobody talked about it. I got a tourist visa to go to India and registered as a refugee with UNHCR.

> I managed to get a place at university in Chennai to study microbiology, but I had no money. To survive, I had to tutor foreign male students. I was at constant risk of sexual harassment and abuse because I was a single woman on my own, so I was a target. When I finished my degree, I went to New Delhi to try to find work. That was when I really felt like I was a *refugee*. I had a degree, but without a work permit I could not get a job in a hospital or pharmacy. I kept asking UNHCR for help, but all they could offer me was backbreaking work making paper plates, for pocket money, in an income generation project. Then someone told me about a training workshop that was being run by some researchers from Australia. I went and that's where I heard about the plan to employ some refugees to set up a community project. I could not believe it when I was employed to be part of it. I never anticipated that I would get a professional job like this: something that would enable me to live with dignity, to grow personally and have some financial security—and that was the same for everyone who worked at RCDP. We all felt that way. But I also had some terrible times in Delhi. I was robbed, attacked and harassed. I felt so vulnerable as a single woman and became overwhelmed with hopelessness and despair. Finally, some hope came when UNHCR told me I would be resettled to the USA. When I was resettled, I told them about what I had done in RCDP and showed them my reference. I arrived in the USA in April and by August I was working as an on-call interpreter. By November, I had a full-time job as a community health worker and I still work there six years later at the Intercultural Mutual Assistance Association (IAAM). I am married now and have three beautiful children. I finally found my mother, whom I feared was dead. She is in Kakuma refugee camp in Kenya and I am trying to sponsor her to the USA. Now I'm planning to enrol in a Master of Social Work. I think about my experience being a refugee in Delhi all the time. It's part of me; it made me who I am today. Without that, I would not do what I do. I would not understand what the refugee women who come here have been through. I can see it in their eyes, even if they don't tell me. It's easier for me to understand them now because of working in RCDP, because of the trainings that I had on human rights and on psychosocial support. Those trainings really opened my eyes and now I see behind the smiles. (Narrated by Hakima October 2021)

After several funding extensions, the project in New Delhi came to an end. We had hoped to secure further funding from a private donor to continue the project, however increasingly complex funding rules for humanitarian and development aid agencies under the newly elected government, prevented this from occurring.

In Nairobi, the urban protection space had become more volatile for refugees, and the project came to an abrupt halt following the 2013 terrorist attack by al-Shabab on Westgate shopping mall. Marcela was caught up in the attack, stranded for several days in a hotel next to Westgate in a terrifying ordeal. Despite this, she oversaw the final evaluation of the program, working remotely with the refugee women. The university did not allow us to return, because of security reasons, and insurers would not cover us. Neither the clinic nor the refugee organisations were eligible to apply for funding on their own behalf.

The Projects were successful because of the flexibility of the Australian Government, who allowed the Projects to move sites, and for the aims and objectives to be adapted to suit circumstances. Many international funders were not flexible and were very reluctant to fund community-based organisations, citing lack of accountability measures as the main reason. The Projects provided rich insights into the major structural, attitudinal, legal and logistical barriers to be overcome before refugees can become self-reliant (Bartolomei, 2018).

Meanwhile, Back in Australia...

In the midst of all of this, at the end of 2013, I officially retired and took on the role of Adjunct Associate Professor. Linda took over as the Director of the Centre for Refugee Research, leading its' transition to the Forced Migration Research Network[2] (FMRN) in 2015. In Australia, we continued to work on various research projects, always with a focus on WaR. Resistance to the arrival of refugees direct to our shores continued, but, under international pressure, and to counter criticism of our draconian border protection policy, the Government increased resettlement numbers, including WaR, and a new category of settlement service provision, targeting complex case management, was introduced to enhance their settlement experience. We had also received ARC funding, with Richard Hugman, for a project entitled '*The Meaning of Rights in Refugee Communities*'. This project built on the previous ARC-funded WaR Projects as well as the unexpected findings from a number of projects undertaken in partnership with settlement service providers.

'Rights Are Ruining Our Lives'

> ***The 'hierachy' of rights***
> This is becoming very common within communities, they are even making fun of it. They are saying... *'Don't you know with this [Australian] government? First, it's my kids, second, it's my wife, third it's my dog - then myself... Things have changed my friend!'*... It is said in a satirical way, but... it's not really [a joke]' (Refugee community member, Brisbane 2014).

Many would assume that, when refugees arrive in Australia after fleeing situations of danger and persecution, their human rights are restored, and as permanent residents of Australia, they are able to access and enjoy their human rights equally with other citizens. Most settlement services based on a human rights framework share this understanding. We also proudly badge our research as human rights-based. It was therefore a shock to realise that many refugee communities were blaming human rights for the problems they were experiencing in settling into their new homeland.

The first major indication had come during the evaluation of the AMES Australia Community Guides project in 2009,[3] when men and women from a number of African communities employed in the program came together to discuss and evaluate its progress. As usual, we began with a human rights training session. The men became angry, shouting that human rights were failing them in Australia. They told about their pain at being unable to fulfil their rights to housing, employment, and basic services, and about racism and lack of respect. They shared their feelings of powerlessness and the loss of their roles as head of their families, and the disintegration of families (Doney et al., 2009).

> *Just imagine 20 years in suffering, then you reach here and face these issues... 'Rights' destroy our families – our women leave us and our children rebel against us.* (Resettled refugee)

This was not only happening in Australia. Similar stories were coming from refugee communities across the world. While it was widely acknowledged that men from patriarchal societies were particularly resistant to women's and children's rights, it was either not taken seriously or seen as too difficult to tackle. Failed projects, addressing men's violence against women, litter the planet. The 'common good' that forms the philosophical basis of the human rights framework is at odds with the cultural and religious values of many refugee communities. Some elders believe that 'human rights' strip away the last vestiges of their cultural heritage, when all they have left in their adopted countries are customs, social norms and family relationships (Khawaja & Milner 2012; Ochala & Mungai, 2016). Women's and children's rights challenge the control men still practice in patriarchal societies as head of the household. Some refugees believed that they had 'the right' to disobey the laws of the country in which they were living, but obeying the law is of course not a choice, and can lead to serious consequences for already traumatised families, such as intervention by child protection.

On the positive side, scholars exploring masculinities had become concerned at what was happening in refugee communities. With a group of long-term partners, we decided to explore this further. In 2013, the ARC funded the project, *The meanings of rights across cultures: an exploration of the interpretation of the human rights framework in Australia*. It covered six sites in three states.[4]

Over three years, consultations and interviews were held with 321 resettled refugees from 26 different countries, to learn about how 'rights' were perceived, experienced, and practised by different groups, and to hear their solutions to the problems. The findings were stark. Women reported a higher level of domestic and family violence than in their home countries, and saw this as a result of the men's reaction to women's and children's rights. This caused tighter restrictions on their movements and even less opportunities for participation. There was also a widespread lack of understanding of what human rights really meant. Not only did many of the refugees not understand what the 'human rights framework' represented, neither did the majority of service providers working with them. The term 'human rights approach' was sprinkled throughout funding applications, but this also was little understood, either by donors nor service providers, and was even less visible in service provision. It was the ultimate 'tick-a-box' exercise. It was also very clear that while human rights training is very popular and often provided in refugee camps, it seldom explored the complexity of rights, nor how they were applied.

The major recommendation from the refugees, both women and men, was that they wanted more information about rights. As an interim response, materials and training packages were produced and widely used by our partners and, for a while, distributed to newly arrived refugees by the Australian Government. However, the problem was in fact much bigger than this. It was based on the socio-legal, economic and political frameworks structuring the experience of refugee lives.

Theory and Discussion

These three very different projects brought together a number of important threads. In each case they highlighted the enormity of the barriers refugees face when they strive to overcome marginalisation and disenfranchisement, and how these are exacerbated by, and reinforce, gender inequality. We struggled at times to tease out answers to problems that somehow lay just beyond our grasp.

Ontological (in)security

At CRR colleague Kristy Ward's suggestion, we used the concept of 'ontological security' to explore the positive impact that genuine participation can have for refugees, by sustaining resilience and well-being, including protection from rape, sexual and gender-based violence (Bartolomei et al., 2018). Ontology can be understood as how we each perceive the reality of our lives, and how that perception influences how we act. 'Ontological security' describes a sense of order and continuity in our experiences. Drawing on the work of R.D. Laing, Anthony Giddens argued that ontological security is the result of having a sense of meaning in one's life (Giddens, 1991). It also involves having a positive view of oneself, the world and the future; it is a good state to be in. However, when our experiences undermine the meaning we have given to our life, it threatens our ontological security. It produces 'ontological insecurity', which throws people into a state of confusion, uncertainty and dread. This is the position in which many refugees find themselves. All sense of order disappears, and it is difficult to imagine a different future. An ontologically insecure person can find it difficult to accept what is happening to them, or plan to change it. Without hope, people wish to return to the certainties they once knew, especially refugees trapped in prolonged refugee situations. It can explain why they cling on to cultural practices more vehemently than those in their countries of origin. Familiar cultural practice may provide the only security they have—even harmful cultural practices such as dowry and forced child marriage, which might have been on the decline in their homeland (Ossome, 2006; Rees & Pease, 2006).

Putting rights and Participation into Community Development

How do you address ontological insecurity? Evidence from the projects in New Delhi and Nairobi highlighted the positive impacts of building resilience, positive identities and feelings of self-worth. This expands to the ability to gain and retain some control over their lives, and is critical to enable people to move forward. In the development and implementation of a genuinely rights-based practice, an emphasis on participation is essential. It challenges the social structures perpetuating social exclusion and disadvantage, and fosters opportunities for the voices of the marginalised to be heard. This includes the recognition that communities have a right to access the resources, skills, rights and knowledge required for effective and sustainable development (Hugman & Bartolomei, 2013; Kenny, 2006).

Community development is increasingly promoted as a long-term protection strategy. UNHCR suggests that this is best achieved through building on the skills, capacity and knowledge of refugee communities to facilitate self-reliance and self-determination (UNHCR, 2009). However, community development practice is frequently divorced from an analysis of the social and political contexts in which it is practised. This is an antithesis to ontological security.

In spite of these challenges, refugees are continuously engaged in the struggle for justice and for the realisation of their human rights, as is evidenced by grassroots refugee women's advocacy organisations across the world—running schools, childcare and women's safe spaces, such as the Women's Madrassa in Nairobi and the women-only picnics in New Delhi (Clark & Copolov 2016; Olivius, 2014). While not a magical panacea, participatory rights-based community development contributes to ontological security, which in turn gives women the confidence to fight for their rights.

Human Rights

Reaching an understanding of the importance of the rights-based approach for community development to be successful, underscored the vital importance of human rights training, as expressed by the refugee communities. Their painful experience of the 'imposition of rights' needed to be explored further. Francesca Klug's characterisation of human rights as 'part law, part philosophy and part political movement' (2000, p. 18) stood out, both in its simplicity and its articulation of the depth of the human rights paradigm. We used this to analyse the factors influencing refugees' perceptions and experiences of human rights. It allowed the complexity of their interactions with the human rights paradigm to be teased out and its different dimensions to be identified more clearly. In reality, these conceptual categories are intersectional and overlapping, and both structure, and are structured by, the society in which they are practised.

With respect to human rights as law, refugees often have a deep mistrust of legal processes and people in positions of legal authority, such as the police and the judiciary. Agents of the law were often perpetrators of violence and persecution in their own countries and in places of refuge and were part of the problem causing refugees to flee. Regarding human rights as philosophy, we have already discussed the values conflict between the conservative, patriarchal cultures of many refugees, and the philosophical roots of human rights discourse (Khawaja & Milner, 2012; Ochala & Mungai, 2016). And finally, the political dimension of human rights: the Refugee Convention came about through a political movement to ensure that refugees have access to human rights in countries of asylum. However, implementing human rights, including refugee rights, is an inherently political act, shaped by political contexts, ideological forces and the pursuit of power. Tragically, rights discourse is often used by those in power to best suit their own political interests (De Varennes, 2006; Glendon, 2001). For many refugees, their experience of human rights has been of a toothless tiger, and they have experienced the failure of the human rights framework when it has been undermined by politics. Despite this, refugees continue to struggle to achieve their human rights, with one significant exception. They want political and civil rights, and many social rights falling into the public sphere. What some reject are the *'private rights', those granted to women and children,* which are perceived as an incursion into the domains of family and religion (Pittaway et al., 2018).

Weaving It Together

We had identified that participatory rights-based community development was a powerful tool for creating a setting in which ontological security could be nurtured and, hopefully, gender equality could be realised and rape, sexual violence and gender-based violence, could be addressed. However, we had also identified that the very premise on which this was predicated, the importance of human rights, was often treated with scepticism and mistrust and rejected by large sections of refugee communities. The next step was to identify ways to take this understanding and translate it into positive action.

Seeking a Way Forward

Unfortunately, there was no formal review process following the Women's Dialogues, and, apart from the promise of the 70th anniversary commemorations in 2021, there would be no opportunity to evaluate the promises made. We continued to fit in trips to the Annual UNHCR/NGO Consultations in Geneva, and the Protection Standing Committee meetings and, with the Australian Refugee Rights Alliance (ARRA) and our trusty UNSW interns, began to informally audit these meetings to see if any of the promises made following the Women's Dialogues in 2011, were being implemented. We usually attended these after trips to the field. This gave us the advantage of bringing issues of concern directly from the grassroots to Geneva. With our partners,[5] we focused on increasing the number of refugee women represented at these meetings. Donors funded outstanding refugee women from ANCORW, the African Women's Advocacy Unit (AWAU)[6] and the Thai–Burma border to participate with ARRA.[7] The 'Australian team' sustained its reputation as effective lobbyists, formidable advocates and a significant presence. Joined by like-minded groups around the world, UNHCR meetings changed character from being sedate and quiet events with predictable outcomes, to a buzz of excitement, with refugee advocates sharing their experiences, and, more importantly, their solutions. We continued to run training for the ARRA group and ANCORW. In the lead up to the UNHCR Global Youth Consultations in 2014, this included training[8] for the Multicultural Young Australia Network (MYAN) because, yet again, the consultation framework was totally gender-silent. As the projects in Nairobi and New Delhi came to an end, we lobbied for increased funding for refugee-led projects, but faced significant resistance to this, with the same old arguments of accountability, bias within refugee communities and the amount of support they needed.

On the upside, we were not alone. Our spirits lifted as the strength of the movement grew internationally. The world-changing work of our long-time mentor, Dr Barbara Harrell-Bond continued, and members of the Southern Refugee Aid Legal Network (SRLAN), continued to raise issues at UNHCR and provide answers in the field. Asia–Pacific Refugee Rights Network, (APRRN), an offshoot of SRLAN, went from strength to strength, and the APRRN Women at Risk Working Group made significant contributions to raising the concerns of refugee women and girls at every opportunity. Our network in the region grew stronger and inspirational refugee women leaders emerged from across the world.

Even the elephants were being challenged, creating a bit of movement in that crowded 'room'. UNHCR was increasingly promoting the use of an age, gender and diversity framework, which responded to gender inequality, the different needs of diverse groups of refugees, addressing rape, sexual and gender-based violence, and the inclusion and participation of refugee women. Some INGOs and NGOs were leading their field in progressive rights-based gender-focused service provision and encouraging others to do the same. Certain faith-based organisations were broadening the interpretation of their doctrines to incorporate gender considerations into their practice, sometimes overtly, at other times covertly. Increasingly, donors were demanding a genuine and proven commitment to a human rights framework, a focus on gender and participation, and an effective response to rape, sexual and gender-based violence, as part of funding conditions.

The silences were being broken about sexual abuse by those in power, positive steps were being taken to prosecute perpetrators and some measure of accountability could be seen in policy and practice, although not nearly enough. Refugee-led organisations were growing in strength and support for them was increasing, but there was still a lot of resistance and much of the rhetoric was tokenistic rather than real. The 'elephant of denial' was still alive and well... albeit becoming increasingly uncomfortable at being challenged.

In the meantime, the push for gender equality was getting much stronger. Refugee women were demanding and attending training courses, implementing programs, and claiming their rights. Things were now happening that were not even dreamed about when we embarked on this journey. The refugee voice was getting louder, and, although the fight was far from won, once you have tasted change you do not let go easily, and refugee women were not letting go. There was so much to take forward, to build on. All we needed was the space. Out of the blue, an opportunity presented itself.

A Policy 'Miracle' and a Major Dancing Moment

In November 2016, US President Obama convened the High-Level Leaders' Summit on Refugees as a side event of the annual meeting of the United Nations General Assembly (UNGA). Due to continuing global concerns over the increasing mass movement of refugees and migrants, a set of commitments to improve the protection of refugees and migrants was agreed upon, as a shared global responsibility, in the New York Declaration for Refugees and Migrants (2016). The Declaration set out a plan to further develop these commitments into what became the Global Compact on Safe, Orderly and Regular Migration *and a* Global Compact on Refugees. It outlined a Comprehensive Refugee Response Framework (CRRF) and tasked the UNHCR to consult widely with all key stakeholders, including academics, to develop a global compact on refugees for adoption at the 2018 UNGA. The New York Declaration commits to continued support for the 1951 *Refugee Convention* and its 1967 Protocol. It included a strong set of commitments to refugee women and girls, including to promote their 'full, equal and meaningful participation' in finding solutions, and a focus on an effective response to sexual and gender-based violence (UN, 2016). It was signed off by all UN Member States. The then UNHCR Deputy High Commissioner for Protection, Volker Turk, declared it a minor miracle.

The game was on again! This was too good an opportunity to miss, providing an opening to influence the most important refugee policy document in years.

Notes

1. In this intensive period, we were joined by Pam Harternick and Sue Cunningham from STARTTS, Oliver White from JRS Sydney and volunteer Jennie Mosely.
2. Initially with Claudia Tazreiter (Co-convenor), Susanne Schmeidl and Caroline Lenette. Now co-convened with Ruth Balint and Sally Baker with additional members including Maree Higgins and Claire Higgins see https://www.unsw.edu.au/arts-design-architecture/our-schools/social-sciences/our-research/research-networks/forced-migration-research-network.
3. Eileen led this evaluation with CRR colleague Geraldine Doney. Research assistance was provided by Tori Vu, Effie Mitchell, Chrisanta Muli, many wonderful interns and committed AMES staff.

4. AMES Victoria, ANCORW, Diversitat, Queensland Program of Assistance to Survivors of Torture and Trauma (QPASTT), NSW Service for the Treatment and Rehabilitation of Torture and Trauma Survivors (STARTTS) and Townsville Multicultural Support Group (TMSG).
5. AMES, ANCORW, AWAU, MYAN, RCOA and SSI.
6. AWAU was established by ANCORW Board Member Juliana Nkrumah with Olivia Wellesley-Cole and Carole Shaw.
7. Several ANCORW board members provided generous donations each year.
8. Linda Bartolomei, Geraldine Doney and Najeeba Wazefadost, a refugee from Afghanistan settled to Australia, delivered the training.

References

Asylum Access. (2014). *Global refugee work rights report 2014*. Asylum Access. Retrieved November 10, 2021, from https://asylumaccess.org/reports/global-refugee-work-rights-report-2014/

Bartolomei, L. Hamidi, M., Mohamed, N., & Ward K., (2016). Refugee community development in New Delhi. *Forced Migration Review*, 48–50, Retrieved November 10, 2021, from http://www.fmreview.org/community-protection/bartolomei-hamidi-mohamud-ward.html

Bartolomei, L., & Ward, K. (2016). *The refugee community development project (RCDP) New Delhi, India, DPP final report*. Centre for Refugee Research, UNSW.

Bartolomei, L., Ward, K., & Garrett, M. (2018). Disruptive right-based community development in protracted urban refugee contexts: The politics of legal recognition. In L. Shevellar, & P. Westoby (Eds.), *The Routledge handbook of community development research* (pp. 23–40). Routledge.

Clark, S., & Copolov, C. (2016, March 7). Refugee-run School in Indonesia a model for government to emulate. *The Conversation*. Retrieved November 10, 2021, from https://theconversation.com/refugee-run-school-in-indonesia-a-model-for-governments-to-emulate-55378

de Varennes, F. (2006). The fallacies in the "Universalism versus cultural relativism" debate in human rights law. *Asia Pacific Journal on Human Rights and the Law, 7*, 67–84.

Doney, G., Pittaway, E., & Vu, T. (2009). *Unsung Heroes: An evaluation of the AMES Community Guides program*, UNSW Centre for Refugee Research, Sydney. Retrieved October 31, 2022 from https://www.arts.unsw.edu.au/sites/default/files/documents/Unsung_Heroes.pdf

Ferris, E. G. (2007). Abuse of power: Sexual exploitation of refugee women and girls. *Signs, 32*(3), 584–591. https://doi.org/10.1086/510338

Freedman, R. (2015, March 20). Leaked report details UN peacekeepers' sexual exploitation and abuse. *The Conversation*. Retrieved November 10, 2021, from https://theconversation.com/leaked-report-details-un-peacekeepers-sexual-exploitation-and-abuse-39004

Gabeau, A. (2017, February 7). Experts warn child refugees being recruited into terrorist groups. *Info Migrants*. Retrieved November 10, 2021, from https://www.infomigrants.net/en/post/1872/experts-warn-child-refugees-being-recruited-into-terrorist-groups

Garrett, M., & Pittaway. E. (2014). *Project respect, building a community-led psychosocial support project focusing on self-care for Somali women and girls in Eastleigh, DPP final report*. Centre for Refugee Research, UNSW.

Giddens, A. (1991). *Modernity and self-identity: Self and society in the late modern age*. Polity Press.

Glendon, M. A. (2001). *A world made new: Eleanor Roosevelt and the Universal Declaration of Human Rights*. Random House.

Hugman, R., & Bartolomei, L. (2013). The ethics of participation in community work practice. In A. K. Larsen, V. Sewpaul, & G. Olin Hole (Eds.), *Participation in community work: International perspectives* (pp. 19–29). Routledge.

Janeway, E. (1980). *Powers of the weak*. Knopf.

Kenny, S. (2006). *Developing communities for the future* (3rd ed.). Thomson.

Khawaja, N., & Milner, K. (2012). Acculturation stress in South Sudanese refugees: Impact on marital relationships. *International Journal of Intercultural Relations, 36*(5), 624–636.

Klug, F. (2000). *Values for a godless age: The history of the Human Rights Act and its political and legal consequences*. Penguin.

OCHA. (2019, May 19). *Gender-based violence: A closer look at the numbers*. UN Office for Coordination of Humanitarian Affairs. Retrieved November 10, 2021, from https://www.unocha.org/story/gender-based-violence-closer-look-numbers

Ochala, Y., & Mungai, N. (2016). The challenges facing single female parents of African background in regional Australia. *Australian Social Work, 69*(3), 1–12.

Olivius, E. (2014). (Un)Governable subjects: The limits of refugee participation in the promotion of gender equality in humanitarian aid. *Journal of Refugee Studies, 27*(1), 42–61.

Pittaway E., Bartolomei, L. A., Pittaway, E. E., Doney, G., (2018). 'Freedom from persecution or continued abuse? An analysis of the meaning of rights in refugee communities', *Journal of Human Rights Practice,10,* 248–267. https://doi.org/10.1093/jhuman/huy015

United Nations High Commissioner for Refugees (UNHCR). (2002, October 11). *Investigation into sexual exploitation of refugees by aid workers in West Africa: Note by the Secretary-General A/57/465*. UNHCR. Retrieved

November 10, 2021, from https://www.unhcr.org/en-au/excom/unhcrannual/3deb32dd4/investigation-sexual-exploitation-refugees-aid-workers-west-africa-note.html

United Nations High Commissioner for Refugees (UNHCR). (2009, September). *UNHCR policy on refugee protection and solutions in urban areas*. UNHCR. Retrieved November 24, 2021, from https://www.unhcr.org/protection/hcdialogue%20/4ab356ab6/unhcr-policy-refugee-protection-solutions-urban-areas.html

United Nations High Commissioner for Refugees (UNHCR). (2021a). *Data*. UNHCR. Retrieved November 10, 2021, from https://www.unhcr.org/data.html

United Nations High Commissioner for Refugees (UNHCR). (2021b). *Learning from experience and seizing opportunities: UNHCR and partner practices in advancing gender equality in Sub-Saharan Africa*. UNHCR. Retrieved November 10, 2021, from https://www.unhcr.org/60901acf4.pdf

United Nations (UN). (2016). *New York Declaration for Refugees and Migrants*. Retrieved October 31, 2022 from https://www.unhcr.org/57e39d987

UNODC. (2017). *Handbook on children recruited and exploited by terrorist and violent extremist groups: The role of the justice system*. United Nations Office on Drugs and Crime. Retrieved November 10, 2021, from https://www.unodc.org/documents/justice-and-prison-reform/Child-Victims/Handbook_on_Children_Recruited_and_Exploited_by_Terrorist_and_Violent_Extremist_Groups_the_Role_of_the_Justice_System.E.pdf

Willis, P. (1978). *Learning to labour: How working-class kids get working-class jobs*. Saxon House, Teakfield.

CHAPTER 12

Refugee Women From the Margins to the Centre—2017–2020

Narrated by Eileen Pittaway

The New York Declaration for Refugees and Migrants 2016—Why Were We So Excited?

Let us explain what made us dance! The outcome of the High-Level UN Summit for Refugees and Migrants hosted by the United Nations General Assembly in New York 2016, was the New York Declaration (NYD). This document is one of the first international policy frameworks to outline the gendered impacts of displacement and migration, and to incorporate significant commitments to address these. After years of advocacy from a multitude of international activists, the clear recognition of the gendered nature of displacement, including the endemic nature of rape, sexual and gender-based violence, was another miracle. Equally important was the recognition of the central role that refugee, and migrant women can, and do, play in developing and implementing solutions. The Declaration sets out a plan to further develop these commitments into what became the Global Compact on Refugees (GCR) and the Global Compact on Safe, Orderly and Regular Migration (GCM). In Annex 1, it outlines a Comprehensive Refugee Response Framework (CRRF) and tasks UNHCR to build on it to develop a Global Compact on Refugees for adoption at the 2018 United Nations General Assembly (UNGA). Most importantly, the NYD commits to continued support for the 1951 *Refugee Convention* and its 1967 Protocol. UNHCR lives in

E. Pittaway and L. A. Bartolomei,
Only Rape! Human Rights and Gender Equality for Refugee Women,
Sustainable Development Goals Series,
https://doi.org/10.1007/978-981-19-0916-0_12

dread that one day, member states will push to rewrite the Convention and weaken it. It now has a new lease of life.

This was the opening we had been looking for, a space to take what we had learned and ensure that it was reflected in the GCR. The aim of the CRRF is to enhance and reform refugee protection and respond to burgeoning refugee numbers. The stated objectives are to: ease the pressures on host countries and communities; enhance refugee self-reliance; expand third-country solutions and support conditions in countries of origin for return in safety and dignity. It calls for a multi-stakeholder approach to refugee protection; expanding to include national and local authorities; international organisations; international financial organisations; civil society; the private sector; the media and refugees themselves. Acknowledging that refugee situations differ in nature it calls for a tailored approach to each movement and a bringing together of humanitarian and development goals. Four designated 'pillars' or areas of concern are: reception and admission, support for immediate and ongoing needs, support for host countries and communities, and durable solutions. The GCR, which UNHCR was tasked to develop, was to contain a Program of Action to respond to these objectives. To our dismay, while the New York Declaration was extremely strong on gender, diversity, responding to sexual and gender-based violence, and the participation of refugees, the CRRF was unacceptably weak. Seriously concerned that gender-related issues would again fall off the table, and not be acknowledged in the GCR, no-one was surprised when we were sucked back into the vortex of international advocacy.

It was easier than we had expected. Volker Turk, then Assistant High Commissioner for Protection, UNHCR, who we respected as a strong and genuine gender champion, visited Sydney in late November to speak at a refugee law conference. At an informal dinner, over a glass of good Australian wine, we expressed our concern that the gender language in the CRRF was not as strong as in the Declaration and asked what he suggested we do about this. He invited us to submit an annotated version of the CRRF, which highlighted places where gender could be more comprehensively included and indicated he would take it from there. (A research associate sitting nearby whispered in an incredulous voice, '*Oh my goodness – is that how international advocacy is done? I thought it was much more formal!*'). We quickly produced the document with suggested amendments to strengthen language regarding the response to refugee women and girls. It was fully referenced to relevant UN documents, academic research and evidence from the INGO sector (Pittaway , 2017), and clearly demonstrated that member states were not being asked to agree to anything new, only to implement previously made commitments. The document was shared with a number of INGOs and disseminated around the world. Early in 2017, UNHCR convened a series of 'thematic meetings' based on the four pillars of the CRRF, at which member states debated and informed the draft GCR. This was a critically important process, because what they decided would determine refugee policy for years to come.

Meanwhile, in the Wider World...

Meanwhile, international tensions were rising even further, contributing to social and economic pressures, and severely affecting much-needed funding to UNHCR. Refugees were fleeing multiple conflicts in the Middle East and in several African nations, and the systematic rape of women in conflict continued to be reported in multiple conflicts, including Tigray and Sudan (Cone, 2021). The major crisis was the plight of Rohingya refugees from Myanmar. A continuation of the decades' long persecution by the Myanmar Government, forced over a million Rohingya to flee their homes in 2016 and 2017. Most fled to neighbouring Bangladesh, already one of the poorest nations in the world. Others escaped to India, Thailand, Malaysia, and other parts of Asia. It was the largest human exodus in Asia since the Vietnam War. Unlike the wealthy European countries, which so fiercely resisted accepting refugees from the Middle East, the Government of Bangladesh allowed the Rohingya to stay, in an area called Cox's Bazar. At the same time, the backlash against immigrants continued across Europe and the USA, as did the rise in right-wing extremism. The NYD was overwhelmingly a response to the increasing disquiet in Europe over the increase in refugees seeking asylum there a concern not extended to those nations in the African continent and Asia who were also receiving huge numbers of forcibly displaced people. In 2016, President Trump was elected to lead the USA. Divisive from the start, many of his policies rolled out in this period were aimed at weakening UN bodies and the moral control they could exert over the USA. He also declared '*war*' on Latino migrants entering the USA via Mexico, many of whom would plausibly have a claim for refugee status.

Impacts on UNHCR and Hopes for the Global Compact on Refugees

The UNHCR sought for new ways to respond as demand for refugee protection grew, budgets shrank, and major donors and member states defaulted on promised donations. The system of protection measures in place when the 1951 Convention was first adopted had been devised for a very different time, when numbers were much lower, and expectations were that the problem would soon be over. Instead, the number of those displaced had increased to 82.4 million by 2020 over one per cent of the world's population. Many of those had been refugees for decades. Resettlement was a mere drop in the ocean, and there was a push to open new complementary pathways, with programs as diverse as citizen groups hosting refugees, tertiary education opportunities, and targeted migration for skilled refugees. They were seeking to widen the pool of potential stakeholders, to include large corporations and funders such as the World Bank, and to a focus on localisation, with local governments and municipalities in host and receiving countries becoming important players in recognising the benefits which refugees could bring. There were calls from some quarters for the Middle East to formally receive refugees, as many had already fled there via Thailand and Bangladesh as 'migrant workers'. A further call was made for improved support to host countries, and for countries of origin recovering from conflict, to enable them to reintegrate their citizens. 'Self-reliance' and 'sustainability' became the buzzwords, and all of this was to be reflected in the GCR. It was imperative that the perspective and needs of all women and girls be included in all of these initiatives.

The Birth of the 'Gender Audits'

The few dollars left at the bottom of our funding bucket were used to travel to Geneva to participate in the UNHCR NGO Consultations in 2017 and for a meeting with the Gender Equality Unit at UNHCR to discuss strategies. We suggested that a 'Gender Rapporteur' be appointed for the thematic meetings to inform the drafting of the GCR, as a good rapporteur can positively influence the outcomes of meetings. There was a precedent, because a gender rapporteur had been appointed for the drafting of the GCM. However, UNHCR felt that this might be a too political move, and if gender got its own rapporteur, then 'all the other *minority* groups will want one too'. Biting our tongues, we then suggested that instead, we conduct a 'gender audit' of the process, recording the gender content of the various meetings and ensuring that it was included in the final outcomes. To our amazement, this idea was enthusiastically accepted, and we were commissioned to undertake a Gender Audit (GA)of the five preparatory thematic meetings and the 2017 High Commissioner's Dialogue (HCD), at which formal debate about the text of the Global Compact was finalised.

The audit was sanctioned and funded by UNHCR's Gender Equality, and Resilience and Solutions divisions. Drawing on our resolve that refugee women should always participate in key processes affecting their lives, the team included women from refugee backgrounds, from the five UN designated regions of the world. They were selected based on their strong commitment to gender equality, to addressing rape, sexual and gender-based violence, for familiarity with the UNHCR policy context, and their rich lived experience. The first team was comprised of Shaza Al Rihawi, from Syria, Apajok Biar, from South Sudan, Tina Dixson, Cheery Zahau, from Chin State, Burma, and Melika Sheik-Eldin, from Eritrea. They were joined at later meetings by Noor Azizah, from Rakhine State, Burma, Andrea Ayala from El Salvador, Najeeba Wazefadost, from Afghanistan and Hafsar Tameesuddin, also from Rakhine State, Burma. They are still working together. From here on referred to as 'the GA team', they are a formidable group of women! They were supported at various times by Geraldine Doney, Effie Mitchell and Charlotte Bell from UNSW and Rachel Tan from the Asia–Pacific Refugee Rights Network (APRRN).

What is a 'Gender Audit'?

The GA team analysed the gender composition of panels and interventions by member states, UNHCR and other speakers, produced 'shadow reports' from each meeting, and compared these with the formal meeting notes. Additionally, they drafted and made interventions in sessions, presented on panels, and provided detailed reports and recommendations after each meeting. Analysis was divided between the process and content and examined how the one influenced the other. As the meetings progressed, there was an increasing demand for the refugee women to participate on panels in side and informal meetings. They consulted widely with their networks and brought current and priority issues of concern to include in their interventions and responses. The women also had access to relevant people within UNHCR Geneva for informal discussions. To ensure consistency and academic rigour, the team arrived in Geneva a full day before each meeting to prepare. We stayed in the same hotel, and worked from six am until after midnight most days, attending breakfast and dinner meetings to prepare, debrief and make sure we made the most of every opportunity to advocate effectively. The team was totally professional, and at the same time passionate in their contributions to the meetings. The women were not paid, but their travel expenses were covered. A number of them took leave without pay from their work to participate.

Our advocacy strategies were carefully designed. Aims and objectives were devised for each meeting. The goal was to generate political will to address the needs of refugee women and girls and unlock the contribution they can make to providing solutions. A key objective was to encourage adherence to the emerging UNHCR Age, Gender and Diversity Framework (AGD) which recognises the different needs of children, youth, women, the elderly, LGBTIQ+ communities, ethnic minorities and people with a disability. These categories are cross-cutting, for example, a disabled elderly woman can also be a member of a minority ethnic group. Pragmatically, it was accepted that we would not get specific mentions of LGBTIQ + communities into the document so decided to use the words 'all women' to ensure their inclusion. We recognised the importance of addressing discriminations experienced by men and including them as crucial supporters of gender equality. The concept of 'intersectionality' was used to explain and shape effective responses. As the gender audits were commissioned by UNHCR, we used language suitable to the UNHCR process and were cognisant of the politics needed to gain the approval of 193 very diverse member states. It was not just an exercise of documentation, but a forensic analysis of content and procedure, drawing on the academic framework developed over the past 30 years, and all our diplomatic skills!

The 2017 Gender Audit of the Drafting Process of the Global Compact on Refugees

The first Gender Audit covered six thematic discussions held between July and December 2017. These were: past and current examples of burden sharing; measures at the onset of large refugee movements; meeting needs and supporting communities; cross-cutting issues; and the High Commissioner's Dialogue (HCD). The process got off to a lacklustre start. Despite a stated commitment to AGD, these were rarely mentioned in the first meetings, and rates of female participation on panels were extremely low. This was noted in the GA team reports to UNHCR and in informal meetings with allies[1] in UNHCR. Thanks to our interventions things changed dramatically by the time of HCD, but it was still an uphill battle. Advocacy was also undertaken to influence key stakeholders before and between the sessions. For instance, an unofficial meeting was convened by the Government of Canada at their Geneva Mission, to develop strategies to support women and girls, and AGD, in the Global Compact. The meeting included 20 member states, UN agencies and NGO representatives, and GA team members. We were cautiously optimistic about what had been achieved in this first official audit.

> *Commitment from the Top* In an intervention on behalf of the GA team in the final session of 2017 High Commissioner's Dialogue, Melika Yassin Sheikh-Eldin stated that we were concerned about the focus on gender mainstreaming as the way of including women and girls into policy and practice. Rather than increasing the inclusion of women and girls, what was happening was that they were being 'mainstreamed into oblivion'. Imagine our delight when in his closing speech, the High Commissioner stated '*We should not mainstream gender into oblivion—it should be exactly the opposite. We take it to heart and want to bring it forward*' (Filippo Grandi, UNHCR, 13 December 2017). This gave us the mandate we needed to continue our advocacy for the inclusion of gender issues in the text of the GCR.

By the end of the year, we had done all we could to ensure that age, gender and diversity, refugee women's participation and a response to rape, sexual and gender-based violence, were comprehensively included in the Global Compact. It was now up to the drafters, and then the member states, who would sign it off at the General Assembly in 2018.

A positive outcome of our participation was a commission to author the World Refugee Council Gender policy paper (Pittaway & Bartolomei 2018). We were also invited to Washington to address key sections of the World Bank about ways in which refugee women and girls could be empowered through targeted gender funding. A third outcome was the determination to seek funding to research the implementation of the gender commitments made in the NYD and the planning for the GCR. Previous experience had taught us that without focused follow-up, promises made were often broken! Unexpected support came from the 'Me Too' movement, which had risen to prominence in the USA in 2017 when a number of women spoke out about their experiences of sexual abuse and sexual harassment in the film industry. It quickly spread, and women around the world disclosed their own experiences. The silence about the extent of rape, sexual and gender-based violence, was cracked, if not broken. This also applied to the abuse of refugee women and girls. It was a strong message that it was not acceptable in any form, anywhere.

2018—A Long Wait for Success

Some good news in March 2018 was that the UNHCR AGD framework had been adopted as policy, and following intensive lobbying led by Jessie Thomson from CARE Canada, it contained UNHCR's Updated Commitments to Women and Girls. The text of the AGD Framework was inspiring!

> ***The Age, Gender and Diversity Policy, and the Updated Commitments to Women and girls*** UNHCR will undertake core actions in the following areas of engagement: 1) AGD-inclusive programming 2. Participation and inclusion. 3. Communication and transparency. 4. Feedback and response. 5. Organisational learning and adaptation. 6. Gender equality and commitments to women and girls.
>
> The updated commitments and core actions to implement them are:
>
> a. Women and girls participate equally and meaningfully in all decision-making, community management and leadership structures, and committees of persons of concern. At a minimum, UNHCR operations will ensure 50 per cent female participants in management and leadership structures under UNHCR's authority, and will advocate the same with partners, including Governments.
>
> b. Women and girls are provided with individual registration and documentation, directly or through support provided by UNHCR. At a minimum, UNHCR will provide women and girls of concern with protection documentation on an individual basis, and will advocate the same with partners, including Governments.

c. Women and girls have equal access to and control over management and provision of food, core-relief items, and cash-based interventions. Depending on the context, UNHCR operations will increase the percentage of women as the primary recipients of assistance within households receiving material and/or cash-based assistance.

d. Women and girls have equal access to economic opportunities, including decent work and quality education and health services. At a minimum, UNHCR will ensure women and girls have equal access to livelihood, education, and health programs it delivers, and will advocate with partners, including Governments, for their equal access to public services.

e. Women and girls have access to comprehensive SGBV prevention and response services. At a minimum, UNHCR operations will adopt and implement SGBV standard operating procedures, operationalising the four main referral pathways for all survivors (safety/security, legal, medical and psychosocial), and will promote the same with partners, including governments (UNHCR 2018, p. 9).

A major omission was the promise to provide sanitary materials. Informally we had been told that across UNHCR's global operations at best one per cent of women and girls had been provided with sanitary materials through UNHCR and it was regarded as an unrealistic and unachievable commitment.

In June 2018, the first-ever Global Summit of Refugees was held in Geneva. It was convened and organized by representatives of eight refugee-led networks including the Australian National Committee of Refugee Women, chaired by Najeeba Wazefadost with support provided by two non-governmental organizations (NGOs), Independent Diplomat and the Refugee Council of Australia. The summit brought together an international group of refugees dedicated to advocacy and inclusion in decision-making, the Global Refugee Network (GRN) was born. Their motto is 'nothing for us, without us'. They are fierce lobbyists for the implementation of the GCR. Paul also lobbied the International Council of Voluntary Agencies (ICVA), who coordinated and presented the official NGO statements at each meeting, to appoint refugees to this role. Slowly but surely, the refugee voice was being heard, and they had a couple of chairs at the decision-making table. Not everyone liked this, because of the loss of power entailed, but the move was underway.

We did not conduct a Gender Audit of the 2018 High Commissioner's Dialogue, as it addressed challenges for Urban Refugees and was not directly related to the GCR. However, the meetings were covered by members of the GA team. After all of the work in 2017, it was extremely disappointing that the agenda and the majority of the session reports were gender-silent. This highlighted the crucial role that the GA team had played in 2017 to ensure that the outcomes of the meeting were gender inclusive. It was a clear indication that the need to articulate gender inclusion was *still* not consistently part of the awareness of either UNHCR senior staff or key stakeholders. Lobbying by a coalition of gender-focused activists on the text of the GCR continued unabated, right up until the draft was presented at the United Nations General Assembly (UNGA) for debate. It was a tense time. While waiting, we pursued our idea of a major research program to evaluate the implementation of gender commitments in the GCR, and to our delight, were awarded funding from the Australian Government to do this. The project, *Refugee Women and Girls, Key to the Global Compact on Refugees* began in September 2018 and is continuing in 2022.

The Global Compact on Refugees

Success!

The GCR was finally adopted on 18th December 2018 at UNGA. The intensive lobbying and advocacy, the amazing work of the women on the team, and the other gender activists and colleagues had succeeded. Right there in the introductory paragraphs are the words:

> Para 13. The program of action is underpinned by a strong partnership and participatory approach, involving refugees and host communities, as well as age, gender, and diversity considerations, including: promoting gender equality and empowering women and girls; ending all forms of sexual and gender-based violence, trafficking in persons, sexual exploitation and abuse, and harmful practices; facilitating the meaningful participation of youth, persons with disabilities and older persons; ensuring the best interests of the child; and combating discrimination. (UNGA 2018)

Major commitments to women and girls include an AGD approach, although the AGD policy was not specifically mentioned, giving states leeway to exclude the LGBTQI + communities. We were still very happy with what had been achieved, and we were thrilled when Gillian Triggs, who took the place of Volker Turk as Assistant High Commissioner for Protection, wrote in the International Journal of Refugee Law:

> The gender audits produced by the University of New South Wales and women with lived experience of displacement... contributed significantly to the inclusion of strong gender commitments in the Global Compact. (Triggs & Wall, 2020, p. 297)

Very importantly, the GCR includes clear accountability measures in a Program of Action to guide states on implementation, follow up and review processes. These include: four-yearly Global Refugee Forums (GRF), with alternate biannual High-Level Officials Meetings (HLOM) to evaluate progress of the implementation of the GCR. The first GRF was scheduled for December 2019, a regular High Commissioner's Dialogue for 2020 and the first HLOM monitoring meeting for 2021, the year of the 70th anniversary of the Refugee Convention. (By the way if you feel you are drowning in acronyms—welcome to our world!).

Despite the gains made, the GCR is a highly aspirational document, and there is a huge distance between the political rhetoric and the reality of refugees on the ground. One of the tragedies of this story is the cavalier manner in which many nations, both developed and developing, discard these obligations and interpret them to suit their narrow political and ideological agendas. At the UNGA, the Trump administration tried to derail the GCR on the grounds that it threatened the sovereignty of the USA, and subsequently refused to sign it (Nichols, 2018). It is testament to the many well-intentioned actors in this process that it now exists, and that all but five member states have signed it.

The Global Forum on Refugees, 2019

When we started writing this book, we thought that the 2019 GRF would be the grand finale, but fate intervened! Welcome to the penultimate UNHCR meeting you will be attending. The aim of the GRF was to ensure that the commitments made in the GCR were implemented, preferably through collaborative effort between diverse stakeholders. The modality was of member states and key stakeholders making pledges or promises of actions and funding to achieve this. The process was supported by sponsorship and co-sponsorship groups established by

UNHCR to encourage collaboration between groups with a similar focus. We lobbied hard to conduct a Gender Audit of these meetings, and after some resistance, but also a lot of support from inside UNHCR, we were again engaged to do this. At our suggestion, part of our role in this round of meetings was to monitor the pledges made for the inclusion of AGD, women and girls and a response to rape, sexual and gender-based violence. Preparatory meetings were held in March, June and November, and the UNHCR/NGO Consultations in July. The Australian mission in Geneva hosted a round table in July, attended by states, INGO allies and UN agencies, to ensure that gender was addressed in each of the sponsorship groups. The GRF was held in December with ten formal sessions, special events, and 24 informal sessions.

It was a very successful forum. UNHCR staff worked tirelessly to achieve often-ground-breaking results relating to AGD. The commissioning of the Gender Audit was a courageous move, as the GA team did not hold back from being critical of omissions in their reports. The support given to the GA team was exemplary and for the first time, they were paid for the work they did as researchers. A wide range of participants from a refugee background were included on panels, as chairs and presenters in formal sessions, and events such as the Women Leaders Roundtable Lunch. The calls on the team to present at events in order to bring a gender perspective were so great that they had difficulty documenting all of the sessions. Fortunately, there were a team of volunteers including from the Global Youth Advisory Council (GYAC) who stepped in to assist. It was an intense and exciting time, adrenalin runs freely at these events, and we had a lot of fun working together sharing the highs and the lows of each meeting and enjoying the excitement of being in Geneva.

There was clear support for the AGD framework and an effective response to rape, sexual and gender- based violence, in the pledging process from member states, from the global south as well as the global north. For example, the Ambassador of Botswana declared that '*rape is a crime! It must come out of the shadows*' (Ambassador of Botswana, GRF 2019). Many INGOs and NGOs previously silent on gender also took a stance on these issues. PLAN International, a major INGO organisation, stated, '*We must look at child and forced marriage as SGBV. It is an institutionalisation of daily rape*'. Three sessions were dedicated to gender issues, one addressing rape, sexual and gender-based violence, one on age, gender and diversity and a third on the engagement and leadership of refugee women and youth. Gender was included in many other sessions including education, energy, health services, and maternal and newborn mortality. The importance of refugee participation was raised across the meetings, but little was said about the additional barriers women and girls face and the support that they might need to participate meaningfully and not just as 'token' women. The (political) failure to address the issues facing the LGBTQI + community, even obliquely, was disappointing, given that they have equal status in international law to all other refugees.

The high quality of the presentations and the analysis of the refugee women on the GA team was commented on by representatives of member states, UNHCR, other UN staff and major donors. They described how it had influenced their awareness of gender issues, because it offered not just 'sad stories' but real solutions.

The power of refugee women representing themselves In the second meeting, one of the team, Tina Dixson, spoke passionately about the need for recognition of the LGBTQI + communities, a taboo topic seldom spoken about at UNHCR meetings. Following this, in the GRF, a representative of a traditionally conservative nation spoke strongly about the need to include LGBTQI + communities. Tina went to thank him afterwards, and he shook her hand, saying, '*How could I not respond after hearing you speak so movingly in the earlier meeting?*'.

The GA team also provided a gender sensitivity awareness raising session to the GYAC team in preparation for the GCR. This led to some impassioned interventions about SGBV from both male and female representatives of the youth team.

The power of a refugee man representing refugee women At a preparatory meeting, a young refugee man from Sudan spoke about the way in which his mother had struggled with sexual abuse, marginalisation and gender-based discrimination. At the end of his intervention, he said 'Look at the audience—it is silent! Everyone is listening to me. I will get applause. And you know why? It is because I am a male talking about gender issues. If I was a female, I would not get this sort of attention, people would be chatting [common in UN meetings]. This is what we have to fix—women have to be heard!'.

Despite these satisfying gains, a shadow dimmed the excitement, when it appeared that commitments to gender equality and the principles of age, gender and diversity were still not consistently championed at senior levels within UNHCR, nor effectively articulated. Their invisibility began with the opening plenary. There were seven male speakers on the panel, with a woman at either end taking notes (like ornamental candlesticks, quipped one GA team member). The High Commissioner failed to include a single mention of age, gender and diversity in his opening remarks. Fortunately, Antonio Guterres, the UN Secretary General and previous UNHCR High Commissioner who opened the event, did mention disability, rape, sexual and gender-based violence, as an issue of concern, but this was the *only* mention of gender made.

It was not only we who complained about the gender imbalance in the opening plenary, and the High Commissioner apologised in the closing session. However, the most demoralising moment was that when as the GA team and other refugee representatives read a powerful and heartfelt closing statement, the High Commissioner and other panellists did not appear to listen. In fact they were caught on video talking among themselves and using phones during the presentation. This lack of courtesy sent a very strong message to the presenters about the value of their participation. It cancelled much of their excitement, satisfaction, and the sense that all of the hours of additional unpaid work, sharing of personal stories and contributions to potential solutions had made a difference. Once again, they were just 'refugees' to be publicly ignored by the very top of the decision-making hierarchy. Whatever the reason for the conversation, it could have been handled much more diplomatically. Many people who attended the final plenary talked about it, and despite the promise made by the High Commissioner at the HCD in 2017, gender had once again been '*mainstreamed into oblivion*'.

Nothing however could contain the excitement of members of the GA team who took the opportunity in that final ceremony to launch a new network, the Global Independent Refugee Women Leaders (GIRWL) network. GIRWL is entirely refugee women-led with a focus on increasing women's participation in local and international policy fora. The announcement noted the important role that membership of the

UNHCR GA Team had made to the formation of GIRWL. To the women's great delight, several INGOs and member states approached them immediately to offer seed funding and support.[4]

Leadership Must Come from the Top

While we knew that there actually was a lot of support in the very senior ranks of UNHCR for age, gender and diversity, we felt that the silence must be broken if UNHCR was serious about these important issues. The silence has a significant effect: it sends direct messages to key stakeholders. Firstly, to those who overtly oppose the inclusion of an AGD approach it vindicates that position by not challenging it. Secondly, to those who are sitting on the fence, it signals that it is not of major importance, they need not act, so they stay with the status quo. And lastly, to those who have been actively supporting the AGD approach, the message is that it is not really a core priority for UNHCR. This was clearly seen in the first evaluation of pledges made by stakeholders at the GRF: only 172 of 1396 pledges directly addressed age, gender and diversity, rape, sexual and gender-based violence, or participation. It was very disappointing. Despite the risk that this could have serious implications for our future relationship with UNHCR, we felt it was important that senior staff become aware of this, so we included it in our final report. The GA Team did not let this spoil their sense of achievement.

Reflections from the refugee women on the value and impact of being supported to participate in the Global Refugee Forum

> I was honoured to highlight the real struggles faced by refugee women related to sexual and gender-based violence across all stages of women's life. I was able to engage, meet and network with many key decision-makers and share my story as well as exchanging information on refugee women-led solutions and share other localised approaches/promising practices by grassroots women in the region. I am very happy that refugee women had the support to attend as there were clear barriers to gender equality, SGBV and women's participation which were still not effectively addressed. (Najeeba Wazefadost 2019)

> Being with the gender audit… has given me a chance to be a small part of the positive changes for women and girls' empowerment, meaningful participation, leadership, inclusion, prevention and response to SGBV and child marriages. It is very encouraging to see how the work of the gender audit team has been recognised by UNHCR and partner organisations… We are the change that we hope to see. (Hafsar Tameesuddin 2019)
> We trained the youth team on gender… All of the participants said it was great to feel they had a purpose in every meeting… international advocacy and auditing helped bring to their attention to how little or how much gender is actually raised in discussions. We have changed lives and developed skills, and have new allies moving forward, which is fantastic! (Apajok Biar 2019)

One Size Does not Fit All—Refugee Women, Key to the Global Compact on Refugees

We apologise that we are not representing these events sequentially, but we though it better to keep the parallel process separate, even though work done in one sphere constantly informed what we did in the other. During the Gender Audit we had been surprised by the number of member states and NGOs who made comments such as these: 'Look, it is not that we are opposed to programs to address age, gender and diversity or sexual and gender-based violence, it is just that we are not sure how to go about it. We hear about principles, and we are shown a couple of examples of good practice, but you know, *one size does not fit all.* Every refugee site is different'. This set us thinking. We wanted to test how some of the commitments might be applied in the field. Building on an ARC research application that we had developed with colleague Caroline Lennette, we devised a project, *Refugee Women and Girls: Key to the Global Compact on Refugees*, to take place over three years in the Asia–Pacific Region. We were very grateful when funding was again provided by the Australian Government[5] The project had two key principles, first an effective response to rape and sexual and gender-based violence, and the real participation of refugee women and girls, and second, a model which responded to the four pillars of the CRRF, and the call for a multi-stakeholder approach. Geraldine Doney was employed as the project manager and subsequently Rachel Tan and Rameen Zahra, Judi Tucker, Anja Wendt and Emma Pittaway joined the research team.

We identified four sites in which to explore the hypothesis that it was possible to devise projects to address the same principles in different places, adapting to local conditions, but adhering to the same intent. It was important that they were places where we had a network of good contacts in both the refugee and NGO communities and UN agencies. They also needed to represent the different protection contexts addressed in the CRRF. We chose Bangladesh, due to the huge numbers of recently arrived Rohingya refugees, Malaysia for its urban refugee population, Thailand as a protracted camp-based refugee situation for whom repatriation and a form of local integration were being discussed, and Burma, working with returnees as part of durable solutions. In Australia,[6] and separately funded, we worked with resettled refugees. Although returned refugees from Burma joined us in some of the early planning meetings, continuing conflict inside Burma meant that we could never enter the country to complete the work there. Among those who joined us in an initial planning meeting in Bangkok, was Akhu from Women for Justice, later killed by the Burmese military in a peaceful demonstration and to whom this book is dedicated.

It was an ambitious project with a team of refugee women, NGO and UN partners in each site.[7] We were also pleased to partner with academic partners Associate Professor Shanthi Thambiah from the University of Malaya, Dr. Sriprapha Petcharamesree and colleagues[8] from Mahidol University, Thailand, and Manzoor Hasan and colleagues[9] from BRAC University in Bangladesh. Very aware of rising resentment from academics from the global south about academics from the global north suggesting partnerships solely to obtain local ethics approval, we worked hard towards building a participatory partnership model through the active involvement in the consultative process, joint project development and shared academic analysis.

In October 2018, we invited potential partners from each site to a three-day meeting in Bangkok. They included refugees who could travel, NGO and UN partners and a representative from the Australian Government. As the Asia–Pacific Refugee Rights Network (APRRN) was a key partner, it was linked to an APRRN meeting. Training on the GCR, and the planning of the following three years' activities were undertaken. A key principle we wanted to investigate was ways to establish a genuinely equal partnership with all the project stakeholders, while acknowledging that this was in fact very difficult. Linda and I, as the senior white Australian academics bringing the funding, had a lot of power. The NGOs had more power than the refugees, the men had more power than the women. The refugees who could speak, read and write English had more power than those who could not. We used the notion of 'knowledge sharing', looking at what each of us could bring to the table to place us on a more equal footing. For example, Linda and I had scholarly learning, had worked a lot with refugees and had contacts at the UN, but we had never been refugees. The refugees had knowledge, experience and analysis of exactly what it was like to be a refugee and solutions for the many challenges they faced. The NGOs had power over resources and the decisions about the types of projects they would offer, and to whom. The frank discussions went well and enabled a firm basis for the project to commence, but it really started me reflecting about the multitude of privileges I held.

While in Bangkok I was asked to run a gender training at the APPRN meeting with our colleague Carolina Gottardo, then CEO of Jesuit Refugee Service (JRS) Australia. The group was mainly men from the Asia–Pacific, all well-educated, many of them lawyers. It was a daunting task. I decided not to start the session talking about gender, but instead about white privilege and the privilege I had there in that room. We then work-shopped the privileges we all had, including male privilege. We listed the benefits that we gained from these, but also the discriminations we suffered from those who had more privileges than us. It was a very moving session, with some very profound outcomes. While over the years, we had acknowledged whiteness theory and the theories addressing the power of privilege, it was not until our work with the GCR that we consciously applied it to our own practice.

Whiteness Theory, Critical Race Theory and Social Privilege Theory

One of the first clear articulations of white privilege is the now famous 1988 essay by Peggy McIntosh, *White privilege and male privilege: a personal account of coming to see correspondences through work in women's studies.* It is still relevant today and has been the basis of subsequent work, which both builds on and critically analyses its usefulness. McIntosh describes white privilege as an invisible package of unearned assets which white people do not acknowledge (McIntosh, 1989). These result in the majority being socially and materially comfortable, confident and oblivious to racial issues. Non-white people, constantly made aware of their non-white status through institutional and personal racism, tend to be alienated and less comfortable. She wrote that most white people are reluctant to acknowledge their privilege and instead look for ways to justify or minimise its effects, believing that their privilege is fully earned. When privileged people do not feel personally powerful, arguments that they benefited from unearned advantages seem unpersuasive and they strongly resist this idea. For us, white privilege provided some answers to the important questions of why it was so difficult to persuade power brokers and key stakeholders to change their positions on the involvement of refugees, in particular refugee women, as critical participants in the decision-making about solutions which determine their lives.

Closely related is the field of critical race theory, which addresses the same issues of systemic racism from a different angle. It studies race and ethnicity rather than whiteness (Burton, 2009; Chen, 2017). The concept of social privilege expands the concept of white privilege to include class, race, gender, sexual orientation, gender identity, citizenship, religion, physical ability, health and age, among others. Intersecting with all conceptualisations of privilege, male privilege and gender inequality are the most dominant as can be seen in the hegemony of patriarchy. Along with race and gender, class privilege is extremely influential. It strongly influences the quality of life, including health, mental health, education, profession, affluence, social status and access to power. White privilege and class privilege are clearly visible in the humanitarian and development arenas, which are dominated by white people and non-white people from the educated ruling class, with males in the majority of management roles. This is a critical barrier to the participation of refugees, especially women, who are seldom either white or upper class.

Privilegeis intersectional A white upper-class man, wealthy and well-educated, is at the top of the privilege ladder. A poor woman from an ethnic minority is near the bottom. In the original ANCORW research in 1989, I analysed the experiences of 320 refugee women from 19 different countries, against social criteria such as education levels, rural or urban backgrounds, literacy and livelihoods in home countries, and ethnicity. It showed there was more similarity between women stratified by social position, than there was between women from the county of origin (Pittaway, 1991).

This thinking has guided our work in the current research cycle. It has not been a comfortable process, and it has forced us to look at the privileges we have brought into the work, and how we have used (or mis-used) them over the years. It is nice to be given space to speak 'on behalf' of others at prestigious international meetings, and have a seat at the policy table. Power is a great ego booster, and to consciously relinquish power can be difficult. The reward for challenging ourselves in this way is a different experience of power, which comes from building trust with individuals and communities and working collaboratively to achieve change. Being conscious of white privilege and other social privileges, and expressing this honestly with colleagues, can at times also be used as an advantage. As we saw right back in the Beijing process, in situations where women activists from the global south have felt it would be unsafe for them to take certain actions or make certain statements in public meetings, they have asked us to do it, knowing that our privilege would protect us. Increasingly now, this is redundant. Instead, refugee women are confronting this assumed privilege and standing up to the forces that oppress them.

The privilegeof human rights In Kuala Lumpur, we organised a forum where refugees presented and explored their analysis and potential solutions with key stakeholders. Representatives from UNHCR and other UN Agencies, some member states, and a number of local NGOs attended and sat in the audience. At their own request, the refugees sat to one side. We had done the human rights training, and made 'human rights necklaces', and we had shared with them the story of the women in Cox's Bazar. In a beautiful bringing together of human rights and privilege, the Rohingya group asked if we could let them have enough beads to make a necklace for all of the people who came to the forum, and they handed these out at registration. When it was their turn to present, they asked that everyone in the room put on their necklaces as the refugees had. They then proceeded to explain what the necklaces represented, that each bead was a right that they were entitled to have, such as safety, food, education and health. At a given moment, every refugee broke their necklace and beads covered the floor. '*Look*' explained the presenter. '*We are refugees. Here are our rights on the floor. But please look around the room. Each one of you who is not a refugee is still wearing your rights*' (Rohingya Refugee Woman 2019).

We incorporated this articulation of privilege into work on the project and the final High Commissioner's Dialogue we audited. At first, we were treated with some scepticism by many of our seasoned non-white partners, but slowly, and while we still make mistakes, it has allowed us to build a level of trust that was not there before. It has confronted local NGOs who have the privilege of class and citizenship to take this into account when working with refugees. It has caused some of the men to challenge themselves and their peers about how they view gender equality. Best of all it has demonstrated respect for the refugees with whom we are partnering.

> The best part of the whole week of training was when we talked about white privilege. We always felt it in our lives, but I have never heard white people say it before… the women still talk about it now. It was so liberating and let us believe we can demand our right to equal, and that we had allies. (Iranian refugee woman, Kuala Lumpur 2019)

In 2019 we travelled to each country to run consultations with the refugee communities and establish the projects. The findings from the consultations were, as predicted, confronting and depressingly similar. Rape, sexual and gender-based violence, was still endemic and the issue which the women most wanted addressed. Stories of horror and pain poured out and the variety of cruelties described by the women was astounding. The women identified issues and suggested many innovative solutions. Most importantly they emphasised their desire to be included and for their skills, knowledge and capacities to be recognised (Fig. 12.1).

> Give us leadership and other trainings… give us information… invite us to all meetings, make sure we are on all committees… visit us…listen to what we have to say …you will see how good we are… (Karen refugee women, Thailand 2019)

Fig. 12.1 Storyboard—refugee women's leadership

The final stage of the project planned for 2020, was to fund small projects that the women had suggested, provide support as required and document the results as part of the GCR evaluation process. Then came the catastrophic change we flagged at the beginning of the book.

Refugee Women from the Margins to the Centre—The COVID-19 Pandemic

COVID-19 was totally unexpected and a disaster that no one was prepared for. As it swept the globe, the impacts on refugee communities were catastrophic. As it entered countries, cities, and camps, humanitarian aid projects were severely curtailed, in many places UN, INGO and local staff were withdrawn from the field, leaving the refugees without traditional supports. Resettlement and emergency evacuations of refugees ceased, leaving many stranded in situations of constant danger, including women at risk. Conditions in camps deteriorated, information was irregular and not always correct. Supplies were difficult to obtain, and health systems already not coping with every day demand had to response to COVID-19-related illness.

And who stepped forward as first responders, pulling together whatever they could, working without funding and performing miracles? Of course, it was refugees themselves, providing information and hand sanitiser, making masks, delivering food and other supplies to isolated refugees, supporting victims of the increased incidence of rape, sexual and gender-based violence, and counselling those who had lost loved ones. No longer 'the beneficiaries' (a demeaning and overused label), refugees became the service providers, playing a key protection role in the maintenance of refugee communities. This should not have been surprising because it was a role they had always played, but when COVID-19 hit and many of the the 'professionals' evacuated the refugee camps, often against their will, it could no longer be ignored. The capacity and ability of refugees, in particular refugee women's groups, was on view in a way it had never happened before. It could not be ignored and was a huge signpost to future ways of service delivery.

We quickly refocused the small refugee-led projects to address the pandemic, and in all three countries refugees took prominent roles. The aims were the same, but the ways in which they were achieved were very different in each country. While 'participation' was the major goal, three different definitions of how it could work were identified. In Thailand, where the women's groups on the Thai Burma Border have been well established, the programs they are running are *entirely refugee-led*, with the minimum support NGO partners. The funding which we were able to provide was auspiced by a progressive INGO who forwarded it on to the Women's Groups, with no cost or conditions. In Malaysia, UNHCR, local NGOs, refugee leaders and community groups have worked together to establish women's social support groups across eight different communities, led by refugee women. They respond to basic needs and provide psychosocial support, especially for those experiencing rape, sexual and gender-based violence. We have named this *a participatory approach with refugee-led programs*. It has established a truly egalitarian method of partnership. In Bangladesh, because of the legal protection framework established by the Government, it was more difficult for refugee organisations to be autonomous, but local NGOs worked closely with the women to develop projects with their input, employing them to run small groups focused on mask-making and providing psychosocial support and information. We have named this model *refugee informed*. These different models have proved that in very different contexts, the same key principles can be implemented in ways which suit the local conditions. *One size does not have to fit all.* We are currently developing these into models of program design, implementation and evaluation, with full training materials.

We are not pretending that the pandemic has not had a devastating impact on refugee communities. Across the world, including in refugee sites, rape, sexual and gender-based violence has risen to the point that it has been labelled a second pandemic (Dlamini, 2021; UNHCR, 2021). But now it is being talked about and refugees themselves are taking action to address it. Refugees have now been acknowledged as key players in service provision, not only vulnerable recipients. This has never before been so universally acknowledged.

From Gender Sensitive to Gender Transformative: The 2020 High Commissioner's Dialogue on Protection and Resilience During Pandemics

Because of COVID-19, there was uncertainty about whether an HCD would even take place. However, in September 2020 it was announced that it would go ahead, and we were approached by the UNHCR Gender Equality Unit to ask if we would be interested in conducting an audit of the meeting. They were delighted to tell us that not only did they not have to fight for this to take place, in fact it was expected it would happen. While the HCD was not nominated in the GCR Program for Action as part of the formal review process; however, we saw it as the first test of whether the commitments made in the document were being implemented by key stakeholders.

The HCD was held as a series of virtual meetings. The nature of the virtual sessions had a major impact, including the function of the GA team. Strategies for completing the audit online were quickly developed. It was extremely testing for the Australian and New Zealand members of the team because the meetings took place in the middle of the night in our part of the world. While the interactive nature of face-to-face meetings was missing, one major advantage of the virtual format was that people from around the world were able to participate, both as panellists and as audience, and questions could be submitted via the web. Many of those posted in advance were answered during the sessions by the panellists. The GA team were omnipresent in the online Q and A chat and submitted well-researched questions in advance for consideration by the presenters. They also presented on several panels.

A Major Thanks to UNHCR—A Magnificent Response

We need not have worried about the reaction of the senior staff[10] to our previous critical report of the GRF. In the 2020 HCD they were inspirational[1]. It was heartening to see that the gender champions in the process were not only female. In his opening address, the High Commissioner Filippo Grandi spoke strongly about what he hoped the meeting would achieve for refugees, internally displaced and stateless persons. Framed in the context of the importance of *trust* to achieve our collective goals, he acknowledged the rise in sexual and gender-based violence during COVID-19, and its impacts, including on women and girls forced to sell sex, and the imperative to address these issues. His analysis was thorough and innovative and, through his strong lead, ensured that a gender lens was applied throughout the Dialogue.

> Those are situations in which gender-based violence unfortunately escalates and has escalated in the pandemic… especially violence against women, against LGBTQI+ communities or individuals.
>
> [After COVID-19] when schools reopen, we don't see the girls coming back. Because meanwhile, poverty has struck so much in their families that for them, the future is bleak. It ranges from forced marriages or exploitation and sometimes survival sex or prostitution. All of these are forms of course of gender-based violence and sometimes it is physical violence and worse. (Grandi, 2020 in Gender Audit Team, 2020)

The High Commissioner constantly referenced the meaningful inclusion of *women and girls* at the decision-making table and included *LGBTIQ+ communities* in his discussion and recommendations. The strong recognition of the role played in the response to the pandemic *by refugee-led organisations, in particular women-led organisations,* and the

[1] Shaza Al Rihawi, Andrea Ayala, Linda Bartolomei, Apajok Biar, Tina Dixson, Eileen Pittaway, Hafsar Tameesuddin, Najeeba Wazefadost (2020) *Combined Gender Audit Report of the Five meetings comprising the 2020 High Commissioners Dialogue on Protection Challenges: Protection and Resilience during pandemics and the 2020 NGO Consultations*, UNSW, Sydney. 60bf7c434.pdf (unhcr.org).

need to provide both space and support was clearly spelled out. He also addressed the need to enable refugee-led organisations to receive funding in their own right. He stated that their place at the decision-making table must become fully recognised and included in all future planning and responses.

In order to address these issues, he outlined the following challenges to the humanitarian and development sector: '*We need specialized activities to address gender-based violence.* [We need] *activities to address prevention and of care for victims and, cash to alleviate poverty. It is important that all organizations set aside resources for that.*'

> Finally, really working with organizations....[by] diversifying partnerships to organizations that are led by women, that are led by LGBTIQ+ individuals, that are led by members of communities that are particularly at risk, ensuring that they are organized... [and] supporting their organization in order for them to be effective. (Grandi, 2020 in Gender Audit Team, 2020)

There was a call for these initiatives to be covered in the next round of pledges, that is, promises made by stakeholders to implement the commitments in the GCR. Throughout the HCD there was also strong leadership on gender and AGD from UNHCR senior staff. Of particular note was the clear articulation of the need to move from a 'gender sensitive' to a 'gender transformative' approach, and what that might entail. A clear recognition of the rights of LGBTIQ+ communities to be included as equals was trail-blazing. There was gender parity on panels and across the Dialogue. The refugee presenters demonstrated a sophisticated analysis of the problems facing refugee communities, and solutions, and discussed the power of community to address these. Organisations and donors were unanimous in praising refugee-led organisations for what they had achieved during the pandemic and promising to change the way things were done in the future. The meeting finished with the challenge for everyone to make operational pledges towards fulfilling these commitments and reporting on them in the next GRF in 2023.

It is tragic that it took such a destructive global event for the contribution of refugee women to be acknowledged and supported. They had been doing many of these things before COVID-19, but it was the withdrawal of international and local aid workers, and the assumption of management and service provision roles by refugees that highlighted their ability and willingness to take on these roles. Without this happening many refugee sites could have descended into chaos. At last it was very apparent that given support and resources they were more than capable of organising their own protection. The GA team documented some very positive pledges, including joint pledges, in which governments, refugee groups, NGOs and civil society joined together to implement commitments and innovative discussions about implementation of the GCR over the next 10 years. The monitoring of the pledges is ongoing, and it won't be until the next GRF in 2023 that we find out what has been implemented.

Notes

1. With thanks for all the support of so many committed UNHCR staff over many years.
2. Najeeba's participation was also enabled through the strong support of Violet Roumeliotis, CEO of Settlement Services International (SSI).
3. Manisha Thomas, WRC, Enzo Tabet Cruz, Plan International, Charlotte Stemmer, Oxfam, Jerome Elie, ICVA, Carolina Gottardo, JRS and Tamara Domicelj, Act for Peace.
4. Global Independent Refugee Women Leaders—World Refugee & Migration Council (wrmcouncil.org).
5. Funding was provided by the Department of Foreign Affairs and Trade's (DFAT) Humanitarian, NGOs and Partnerships Division and Gender Foreign Policy and Asia Engagement Section.
6. Australian partners include AMES, ANCORW, Diversitat, MDA, MYAN, SSI and TMSG.
7. In Bangladesh partners include Rohingya refugee women leaders, Relief International, UNHCR, IOM, Danish Refugee Council, Action Aid, tai, NGO Forum for Public Health, Lighthouse and UNICEF. In Malaysia partners include refugee women leaders from Afghanistan, Burma, Pakistan, Somalia, Sudan and Yemen and NGO partners Tenaganita, the Malaysian Social Research Institute, Asylum Access and UNHCR. In Thailand partners include The Karen Women's Organisation, The Karenni Women's Organisation, The Muslim Women's Organisation and Association, Women's League of Burma and NGO partner The Border Consortium (TBC) and UNHCR.
8. Ratawit Ouaprachanon and Dr Mark Capaldi also joined the academic team in Thailand.
9. Nahida Akter, Muhammad Badiuzzaman and Dr Monira Ahsan joined the academic team for part of the work in Bangladesh. Tasnuva Ahmad joined the team in 2022.

10. We congratulate High Commissioner Filippo Grandi, Assistant High Commissioner Gillian Triggs, Deputy High Commissioner Kelly Clements, and Mr Andrew Harper, special advisor on Climate Action for their frank analysis of gender and SGBV as central to the debate.[1].

References

Burton, D. (2009). Non-white readings of whiteness. *Consumption Markets & Culture, 12*(4), 349–372. https://doi.org/10.1080/10253860903204535

Chen, J. M. (2017). The contentious field of whiteness studies. *Journal for Social Thought, 2*(1), 15–27.

Cone, D. (2021, August 26). Women in tigray face increased risk of sexual exploitation and abuse. *Refugees International*. Retrieved November 10, 2021, from https://www.refugeesinternational.org/reports/2021/8/23/women-in-tigray-face-increased-risk-of-sexual-exploitation-and-abuse

Dlamini, N. J. (2021). Gender-Based Violence, Twin Pandemic to COVID-19. *Critical Sociology, 47*(4–5), 583–590. https://doi.org/10.1177/0896920520975465

Gender Audit Team. (2020). *Combined Gender Audit: Gender Audit Report of the Five meetings comprising the 2020 High Commissioners Dialogue on Protection Challenges: Protection and Resilience during pandemics and the 2020 NGO Consultations*, UNSW, Sydney. Retrieved October 31, 2022, from https://www.unhcr.org/60bf7c434.pdf

McIntosh, P. (1989). *White privilege: Unpacking the invisible knapsack*. Retrieved November 19, 2021, from https://psychology.umbc.edu/files/2016/10/White-Privilege_McIntosh-1989.pdf

Nichols, M. (2018, November 14). U.S. alone in opposing U.N. refugee text over sovereignty concerns. *Reuters*. Retrieved November 10, 2021, from https://www.reuters.com/article/us-usa-un-refugees-idUSKCN1NI2VB

Pittaway, E. (1991). *Refugee women: Still at risk in Australia*. Bureau of Immigration Research (BIR).

Pittaway, E., & Bartolomei, L. (2018). From rhetoric to reality: Achieving gender equality for refugee women and girls. *World Refugee Council Research Paper Series No. 3*. Centre for International Governance Innovation, Ottawa Canada. Retrieved November 10, 2021, from https://www.cigionline.org/publications/rhetoric-reality-achieving-gender-equality-refugee-women-and-girls

Pittaway E., Bell C., & Bartolomei, L. A. (2017). Refugee women and girls in the comprehensive refugee response framework. *Forced Migration Research*

UNHCR. (2021). *COVID-19 pandemic worsening gender inequalities for refugee women and girls, UNHCR*. Retrieved October 31, 2022, from https://www.unhcr.org/neu/52701-covid-19-pandemic-worsening-gender-inequalities-for-refugee-women-and-girls.html

CHAPTER 13

Epilogue—The Move from Rhetoric to Reality—2021–2022

It Was Quite a Rollercoaster Ride

In Part 1, you were invited to join us on the journey to have the prevention of rape and sexual abuse of women enshrined in law and policy—and we succeeded. In Part 2, we engaged with refugee women in the fight to implement these and to achieve 'participation', which in this context is the involvement of refugee women in decision-making in all aspects of their lives. Our goals widened, from addressing rape, sexual and gender-based violence to levelling the gender equality playing field. The inclusion of refugee women and men has slowly moved beyond tokenistic and now acknowledges their strengths, wisdom and experience. The notion of 'mainstreaming' women, girls and gender into oblivion has been transformed into articulation of their different needs and capacities. With refugee activists and colleagues from around the world, we have contributed to a series of gradual changes, which laid the foundation for the recognition of the capacity of refugees and the contributions they can make to International Protection. These were finally recognised when they took a lead role as 'first responders', providing assistance and support to their communities in the COVID-19 pandemic. Refugee women have

E. Pittaway and L. A. Bartolomei,
Only Rape! Human Rights and Gender Equality for Refugee Women,
Sustainable Development Goals Series,
https://doi.org/10.1007/978-981-19-0916-0_13

now taken a lead, and UNHCR and stakeholders around the world are responding.

We leave the last words of the story to refugee women and project partners in our Project in Malaysia, several who have spoken at virtual meetings of UNHCR Geneva and at the Generation Equality Forum in Paris, 2021. We joined them for a virtual celebration of the 30th anniversary of Tenaganita, one of the project partners and were moved to tears by their amazing analysis. Here are the voices of just some of the presenters.

Deborah Biak Tha Bor, the Myanmar Ethnic Women Refugee Organisation (Burma)

We are a group of women who have a passion for women's empowerment and rights, and address various issues and needs happening to Myanmar refugee women in Malaysia. We do this through the provision of services in such areas such as socio-economy, empowerment, sexual and gender-based violence, women's rights, security, family issues, family health and physical health and mental health as well as successful reproductive health. I am one of the focal points who facilitate in the working group which stands for our women and girls at risk. They share with us their troubles and burdens, we are always listening to them very attentively and we are trying to encourage them, 'you are not alone here, we are always here for you, to hug you and to get out of that crisis'.

Leena Al-Mujahed, Founder, Yemeni Refugee Women's Organisation

This is the first project where I can see a woman leader from the different communities, different language, different organisations, different religion, everything is different, but we come together here to support each other and to share all the things that we need for our community and how to help our community. So, you know that the full realisation of women in decision making is very important, and this is what we are doing during this project. It is amazing that we have, as women leaders, we have the decision making, we have the space to say what we want to do, what is our opinion, what are our future plans, what is the project that we need. Always we find the support, that everyone on this project brings from NGOs or Tenaganita or UNSW. These projects brings us gender equality, leadership, decision making, participation of woman and girls and humanisation.

Fajar Santoadi, Project Manager/Counsellor, Tenaganita

Sometime in the past refugees had bad treatment or bad experience with organisations. [Through this project] with partners we now have a tool, so this platform has become a classroom for us to really make use of our privilege for the goodness of others. We are now learning together, and it is a participatory approach in a real sense that we are all trying to make it a reality. [I was disillusioned] because before I have attended many consultations involving so many stakeholders, and after we finish, we come back with problems and then the next action is not always clear [nothing happens]. What I have experienced in the last three years is really different. I have been seeing the most important stakeholders, which are refugee woman leaders, that have been participating and then not only telling the problems but planning to overcome the problems. It is not an empty promise, I have overcome that fear of 'only consultations' because many of us have been willing to be involved in this messy *process, back and forth, and sometimes difficult, but also what is important is trusting each other, trusting woman leaders, trusting whoever involved in this. So, I am going to just end by saying of really feeling joy and pride that everyone has been so trusting and so believing in each other in this participatory approach.*

Hui Ying, Executive Director, Asylum Access

This project taught me about risks and about not being afraid to fail. It allows space to learn to work together, to be more collaborative and creative. So, I do think about meaningful participation - it requires us to be flexible and more importantly, of being in continuous and open communication to build that kind of trust. What does it mean to be inclusive, what does meaningful participation mean? I think I got lucky that in my formative years I came across this project - I was like wow! Okay, this is what it means to be participatory. Since then, it has had a huge impact on the way that I think about the way we do work, about our partnership, about the way we think about our programs.

Joshua Teh, Protection Associate, Child Protection & Gender-Based Violence Unit, UNHCR

So, the project always has been centred around refugee women and human rights and there for me is the essence of meaningful participation. It really has been a privilege to work together with each and every one of you women leaders and it is so important that when we say that this is a partnership, it isn't merely lip-service, so for the women leaders: you are wonderful, you are committed, you are strong and you know best what your needs are, what your community needs are and it was important that we learn from you. You provided input on session modalities, projects, accountability mechanisms and the extent of sexual violence. It was crucial for the focal points of your group members to know that they were not just there to rubber-stamp UNHCR or UNSW or NGO partners, you were in fact crucial and important, and this project could not have run without you. For UNHCR I think that the level of trust that has been developed, the open, direct communication and the way all the partners have worked together, really provides us with a blueprint on projects moving forward, their aim and meeting refugee needs. It enables us to learn more about refugee experiences, your triumphs, your challenges and I think more importantly, it ensures that refugees around whom our work is centred, know that they are being heard, that you have a voice and that your efforts are truly crucial in bringing change to your community.

So, this is where we leave you, feeling more positive about significant change than ever before. Rape, sexual and gender-based violence, against refugee women and girls, have not stopped, but it is acknowledged and seen in the broader context of universal gender equality. Refugee women have proven time and time again that they are resilient, capable, knowledgeable, unafraid and strong. They themselves are part of the fight for security and justice.

They now stand on a level playing field with other women and are fighting for their rights. The genuine recognition of the value of the work by refugee women-led groups and organisations during the COVID-19 pandemic, often as first responders, supporting both their own and host communities, has been seen and heard. Moves are underway to secure funding for their work. Refugees are now firmly in the centre of the advocacy circle, and we take our place on the support team (Figs. 13.1 and 13.2). The application of an AGD approach which recognises the different needs of women, men, people with disability, the LGBTQI +

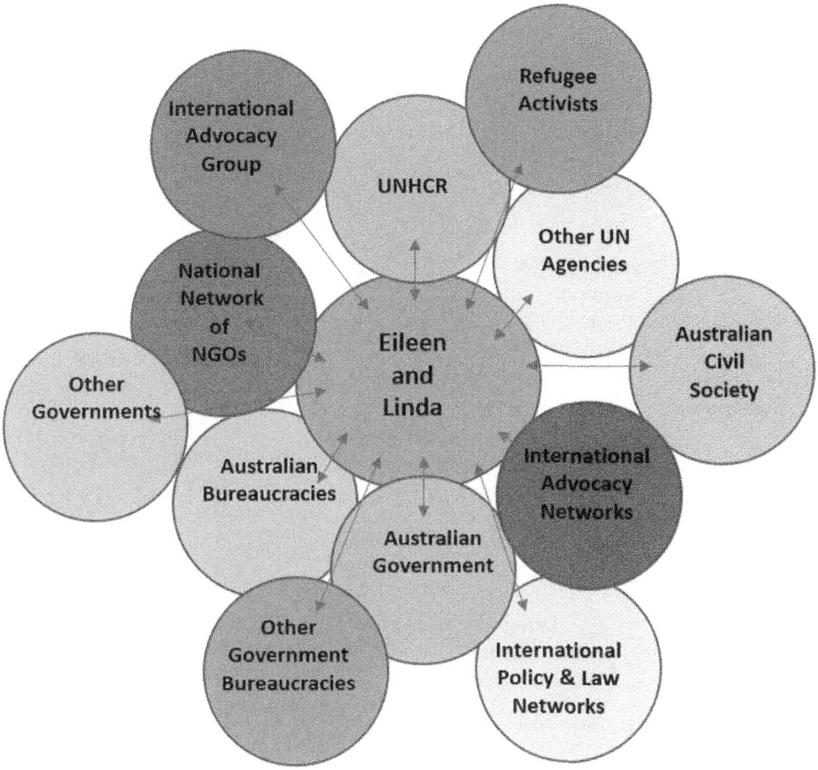

Fig. 13.1 Profane View

communities and other minorities is becoming more common. The call by the UNHCR High Commissioner to move from 'Gender Sensitive' as a guiding principle of the work of UNHCR and its partners, to that of 'Gender Transformative' has the potential to bring about major change.

The participation of refugee women and men from the local to the global has moved from tokenistic and their strengths and knowledge are now acknowledged. The notion of 'Mainstreaming' women, girls and gender into oblivion is slowly being transformed into a genuine acknowledgement of their different needs and capacities. Whether this translates to significant action on the ground is the challenge for of future activists, both refugees and their supporters.

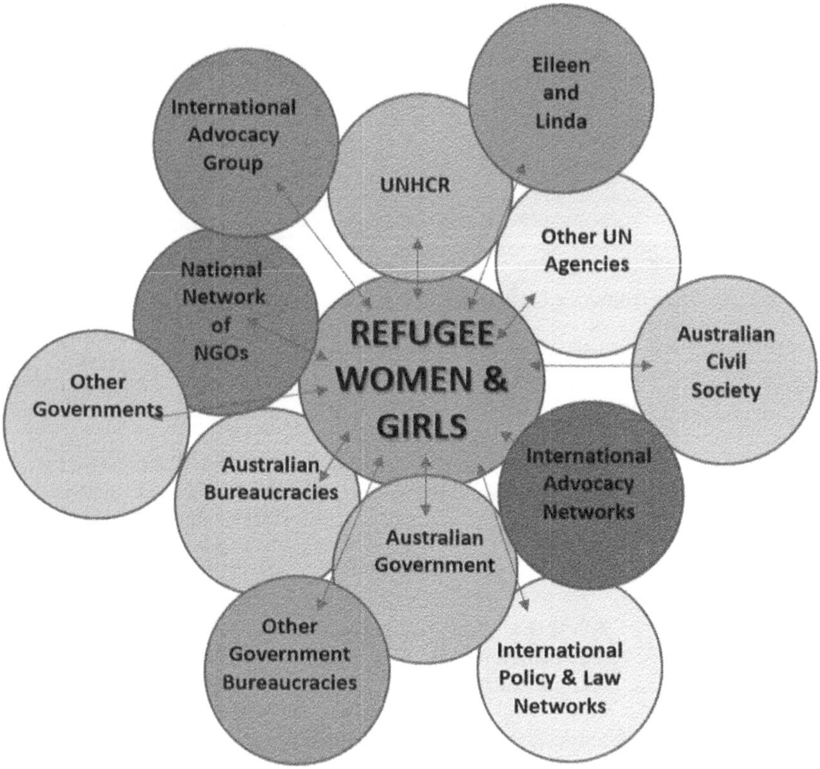

Fig. 13.2 Profane View Updated

We have celebrated many successes over the years, in sometimes small and sometimes large ways. Our work has contributed to the recognition of rape as a war crime, a crime against humanity, and in some instances an act of genocide in the Beijing Platform for Action, and then in the 1998 Rome Statute for the International Criminal Court. The Women's International League of Peace and Freedom (WILPF) must be acknowledged for their ground-breaking advocacy to achieve the first of ten Security Council Resolutions on Women, Peace and Security (WPS) including Security Council Resolutions 1325 (2000) and 1820 (2008). We prepared the first draft of the UNHCR soft law Conclusion on Refugee Women at Risk (2006) and lobbied until it was accepted. A

Heightened Risk Identification Tool (HRIT) which was developed as part of one of our research projects was adopted by UNHCR. In 2010/2011, seven Women's Dialogues were conducted with refugee women across the world. Women representatives from those meetings presented their findings in Geneva as a part of the celebration of the 60th anniversary of the 1951 Refugee Convention, with very positive policy and funding outcomes. The New York Declaration, 2016, and the Global Compact on Refugees, 2018, provided many opportunities to continue the fight and see solid commitments to refugee women and girls. A Gender Audit process undertaken with refugee women at all key UNHCR meetings related to the GCR and its implementation has informed the ongoing pledging process from member states and other key stakeholders. The Women at Risk Program is still available to refugee women in very vulnerable situations, and Australia was among several states who particularly responded to women and girls as a priority for evacuation from Afghanistan following the fall of the Government in August 2021.

Most important of all we have seen the inclusion of refugees as participants in UNHCR Forums and at the decision-making table—a process which is evolving and growing as we write. Refugee representatives have gone from not attending meetings, to attending but being silent. Next came 'token' speaking spots—refugees telling sad stories. Now, they are respected for their experience and analysis and invited to be on panels and to chair meetings. This has happened because of their capabilities and determination, and through the work behind the scenes of a multitude of refugee representatives who made this happen. Vibrant refugee networks, such as the Global Refugee led Network (GRN), the Asia–Pacific Network of Refugees (APNOR) and Global Independent Refugee Women Leaders (GIRWL) all with strong commitments to human rights, gender transformative and inclusive AGD approaches, are taking the lead in advocacy and work on the ground. Their lobbying was validated by the recognition of the effectiveness of refugee women and women's groups as first responders in the COVID-19 crisis. They have moved from invisible to omni-present.

Throughout the book, we have endeavoured to demonstrate how theory and analysis have provided intellectual tools which have enabled us to move forward. The gradual building of the framework, which includes strong political and ideological analysis using critical theory, has enabled us to comprehend different versions of reality and the complexity of social structures. Feminist theory was a guiding light throughout. This provided the tools we used for unpacking the meaning of human rights, our driving principle, to the many actors involved. It illustrated the power of genuine participatory rights-based community development. The importance of the many identities that we all have and the discrimination and oppression which come with many of those was a big lesson. The impact of intersecting discriminations on the ability of people to access human rights and resilience gave us a tool to shape policy and service responses. The value of honesty and a willingness to face our own privileges, of using the power we have for the good, was a lesson learned hard. Over the years, this framework has guided us in developing the *Reciprocal Research* methodology and our work on the ethics of research with refugees. We are proud and privileged to have been part of this movement. We have amazing refugee companions to walk with now, along with the activists, organisations and others who have been integral to this story, of which we have been but a small part. We hope that the theoretical framework we have produced will be of assistance to others in future. The crazy roller-coaster ride did not beat us and our dancing shoes have holes in them from all the wonderful moments we have shared.

The Silences Are Broken, the Women Are Recognised

As a fitting end to the book, we were delighted in November 2021 that the UNHCR NGO Innovation Award 2022, which celebrates innovative approaches for protection and delivery of services to refugees and other persons of concern, focuses on women-led and girl-led organisations. At the same time, the High Commissioner released a poster for the 16 days of activism against Gender-Based Violence (Fig. 13.3) and (Fig. 13.4).

Postscript May 2022

It is with a mixture of great sadness, fury and despair that we write this short postscript. We could not let this book go to print without acknowledging the horrendous suffering and human rights abuses being experienced by so many women and girls in the current invasion of Ukraine by Russia.

As the news of the conflict reached us, we looked at each other and said: 'How long before the rapes begin?'. It was less than two weeks before the reports started to emerge. Widespread rape and sexual abuse

Fig. 13.3 UNCHR NGO Innovation Award 2022

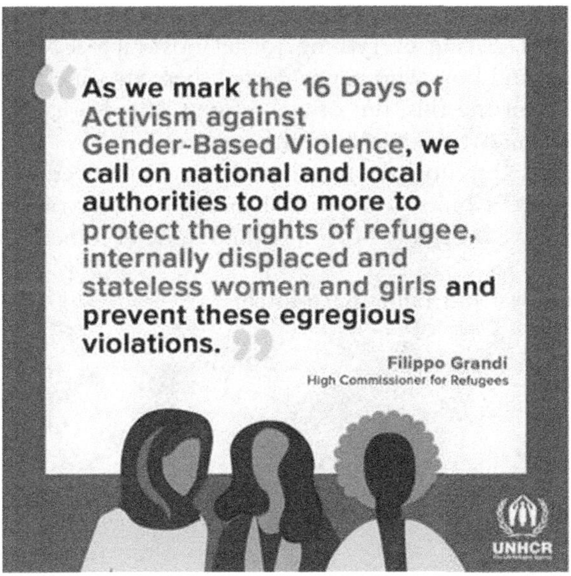

Fig. 13.4 16 Days of Activism against gender based violence poster

of women and girls, and now reports of rape of men and boys. Young women desperately seeking 'morning after' pills, knowing full well that they were likely to be raped. Then, the credible and confirmed accounts of the groups of girls and women held for days, and continually raped by soldiers, being told that they were deliberately impregnating them, so that they would never want to have sex again nor give birth to Ukrainian children. Girls as young as 14 pregnant. Women being murdered after being gang raped or taking their own lives.

What does all this tell us? That rape is still an organised strategy of war. That women are still regarded as 'chattels', as sub-humans by some, and that when given permission and the knowledge of impunity, many soldiers are happy to comply.

Rape is not a weapon of war, it is a physical act that is weaponised. Can we ever change this? Has 33 years of work just been a waste of time? Some things have improved. Rape in conflict and refugee situations is now being widely reported. Women are denouncing rapists and demanding international protection and justice. Good men are speaking out against the men who do it and the men who give the orders to use it as a strategy. Some aid agencies are risking everything to get post-rape services to women and girls, men and boys who have suffered these atrocities. Many governments are supporting this, but others refuse to fund organisations which provide sexual and reproductive health services.

We will not give up, but the fight is far from over. Human rights activists, refugee women and all stakeholders must continue to work together until we stop this horror. It must not become a never-ending story.

Eileen Pittaway and Linda Bartolomei

Correction to: Anger and Despair—Naming the 'Elephants in the Room'—2012–2016

Correction to:
Chapter 11 in: E. Pittaway and L. A. Bartolomei,
***Only Rape! Human Rights and Gender Equality for Refugee Women*, Sustainable Development Goals Series**
https://doi.org/10.1007/978-981-19-0916-0_11

The original version of this book was inadvertently published with an incorrect author name in Chapter 11, which has now been corrected. The book has been updated with the change.

The updated version of this chapter can be found at
https://doi.org/10.1007/978-981-19-0916-0_11

E. Pittaway and L. A. Bartolomei,
Only Rape! Human Rights and Gender Equality for Refugee Women,
Sustainable Development Goals Series,
https://doi.org/10.1007/978-981-19-0916-0_14

Index

E. Pittaway and L. A. Bartolomei,
Only Rape! Human Rights and Gender Equality for Refugee Women, Sustainable Development Goals Series,
https://doi.org/10.1007/978-981-19-0916-0

The manufacturer's authorised representative in the EU is Springer Nature Customer Service Centre GmbH, Europaplatz 3, 69115 Heidelberg, Germany. If you have any concerns regarding our products, please contact ProductSafety@springernature.com

Printed and bound by CPI Group (UK) Ltd, Croydon, CR0 4YY
30/09/2026
02225327-0010